Transport Planning
Second Edition

Transport, Development and Sustainability

Series editor: David Banister, Professor of Transport Planning, University College London

David Banister

Transport Planning

Second Edition

London and New York

First published 1994 by E & FN Spon
Second edition published 2002 by Spon Press
11 New Fetter Lane, London EC4P 4EE
Simultaneously published in the USA and Canada
by Spon Press, 29 West 35th Street, New York, NY 10001

Spon Press is an imprint of the Taylor & Francis Group

Typeset in Sabon and Imago by NP Design & Print, Wallingford, Oxfordshire
Printed and bound in Great Britain by TJ International Ltd, Padstow, Cornwall

British Library Cataloguing in Publication Data
A catalogue record for this book is available from the British Library

Library of Congress Cataloging in Publication Data
A catalog record for this book has been requested

ISBN 0-415-26171-6 (hb)   0-415-26172-4 (pb)

Contents

> The dominant economic fact of our age is the development not of manufacturing but of the transport industries. It is these which are growing most rapidly in volume and in individual power. (Alfred Marshall, 1890)

One of the most frequent questions raised by my students has been what book would I recommend. My answer over the last 10 years has been consistent, namely that I cannot recommend one book. Several books are useful, but none seems to present a well argued analysis of how transport planning evolved, what are its strengths and weaknesses, what it has achieved, how it relates to actual policy decisions, and where it is likely to go in the future. In this revised edition of *Transport Planning*, I have attempted to address all these issues and so to fill this gap in the literature.

However, the book does not attempt to cover all aspects of transport planning, but concentrates on the processes by which analysis links with policy and the changes which have taken place over the recent past. Most emphasis is placed on passenger transport (public and private) and on land-based modes. Although the context taken and the perspective given is from Great Britain, extensive coverage is presented of experience in Europe, the United States and elsewhere to establish whether lessons can be learnt. The importance of the issues raised by transport planning is not restricted to one country, or to one policy-making context, or to one analytical approach. Cross cultural analysis allows very different attitudes and perspectives to be brought together on common problems, and increasingly solutions are seen as being international, with analysis tools also being transferable between countries and cities. Attention is also given to the new transport technologies, the environment, and important policy issues such as market and planning failures. Case study material is taken from all scales, from the local to the citywide, from regional to national, and from one country to another.

The book is divided into in two main parts. The first is a *Retrospective Analysis* of the development of transport planning as an important and legitimate area of research and practice, together with its rejection and re-emergence. Experience here is taken from Great Britain and the United States, and direct parallels are drawn with the similar developments in planning analysis. Links are also drawn with evaluation methods in transport and planning, and the instrumental role which they often played in several of the major transport decisions in the 1970s. There is a review of the radical policy alternative presented by the market approach to transport provision, with reductions in public expenditure, regulatory reforms, and the reduction in state ownership of transport enterprises. This radical agenda is complemented by a new chapter on events since 1997, with a change in UK government and in direction, at least in principle, but the difficulties of finding 'the third way' have proved problematical in transport policy terms. The first part ends with a discussion of the limitations of transport planning in terms of both its theoretical foundations and of the approaches being developed. It is argued that new and modified methods are required to investigate the main elements of the current policy agenda. These new approaches are available, but need to be combined with a more proactive stance from transport planners themselves. This new approach requires more openness and the involvement of all interested parties in that process. This is the new inclusive transport planning that is investigated in the second part of the book.

First, there is a *Comparative Analysis* of experience in three European countries (Germany, France and the Netherlands) and the United States. This covers both transport planning and evaluation, where in each case, different approaches have been developed in response to national policy issues and the different cultural and analytical traditions. This experience provides a useful context for interpreting the new agenda in the United Kingdom. Secondly, there is a more *Prospective Analysis* of the key issues facing transport planners in the twenty-first century. Certain contextual changes are all likely to have a significant effect on transport demand, and there is a new imperative to replace existing infrastructure, to build new infrastructure, and to ensure the optimal use of existing infrastructure. All these issues will necessitate different analytical approaches, particularly where new forms of financing are required. The role for transport planning is likely to change as the new political relationships between the state and market are stabilized. Some form of strategic vision is required together with a planning framework within which the market can operate. The role of the transport planner will be to input to this process at a variety of scales, to establish a clearer understanding of the links between longer-term economic growth and the shorter-term fluctuations in the monetary economy, and transport demand; to act as a promoter of economic development; to ensure that the city regains and maintains its position as an attractive place in which to live and work; and to provide assistance where the market fails, in particular by meeting social needs. As Buckminster Fuller prophetically stated, 'Hope in the future is rooted in the memory of the past, for without memory there is no history and no knowledge'.

David Banister
March 2002

Acknowledgements

Many individuals have helped in bringing this book together, and I would like to thank them all. They include colleagues at University College London, in particular Ian Cullen (now with Investment Property Databank) and Roger Mackett, and others from many European research centres. Alain Bieber and Peter Nijkamp commented on Chapter 5, and Peter Hall has provided an invaluable sounding board for ideas. He was also prepared to read the whole manuscript in draft. Others have helped with the production of the text and diagrams – Lizzie, Paulo, Ann and Jennifer. The research was funded by the UK Economic and Social Research Council as part of their Transport Research Initiative (1987–1991).

Since then, I have benefited from learning more about the complexity of transport planning and have had many inputs as to what should be in the second edition of the book. In particular, the development of European research collaboration has resulted in a better understanding of different transport and non-transport perspectives – thanks to colleagues in the POSSUM, DANTE, SAMI and TRANSPLUS projects among others, as well as researchers in the many networks, including NECTAR and STELLA. I hope that this new edition does justice to their thoughts and the debates of the last eight years.

Lizzie, Alexandria, Florence and Sally

Chapter 1

Transport and travel in the last 40 years

The era of the car

The car has revolutionized the way in which we look at travel and communications. Before the advent of mass car ownership in the 1960s, people travelled short distances by foot or bicycle, while longer journeys were made by bus or, occasionally, by rail. Life was centred on the locality in which one lived, with work, schools, shops and all other facilities being available locally. Travel outside the community was only undertaken for special reasons such as visiting relatives or going on holiday. In 1960, each person travelled on average some 5,600 km. By 2000 that figure had increased to nearly 11,000 km (Table 1.1). No one thinks twice about using the car to travel to work, to do the shopping, or to take the children to school. It is used as a form of transport, as a local bus service, as a goods vehicle, and as a statement of status. It is often the second largest single item of family expenditure after the home and is treated with respect and affection. It is generally regarded as the most desirable form of transport and will normally be used, no matter how attractive the alternatives might be. The user can always find a reason why the car is necessary for a particular journey.

Subjectively, these arguments are clear, but objectively the dominant preference for the car is (or should be) less clear. Cars are a major cause of premature death in all Western countries. In Britain 3,400 people are killed on the roads each year, with a further 43,000 people seriously injured (1998), yet this is a cost that society seems prepared to accept. Transport is a major user of energy, with all forms of transport consuming some 34% of total energy consumption in Britain. Transport is a major contributor to environmental pollution and is one sector where most of the trends are in the wrong direction, with increases in emissions of greenhouse gases (some 26% of carbon dioxide), contribution to acid rain (54% of nitrogen oxides and small amounts of sulphur dioxide) and other gases which have effects on morbidity, fertility and mental development (57% of lead, some 74% of carbon monoxide, and 28% of particulate matter – PM_{10}).

The total amount of travel by road (measured in billion passenger-kilometres) has increased in the last 30 years by almost three times with only 8% of that figure attributable to population increase. The remainder is due to increases in mobility, which relates to the amount of travel in terms of the numbers of journeys and the lengths of those journeys (Table 1.1). Whatever the criteria used, the result is the same, namely the relentless growth in the amount of travel. The 1960s had increases of about 120 billion passenger-kilometres of road travel, while the 1970s had a similar level of growth of 120 billion passenger-kilometres. But in the 1980s there was an increase of nearly 150 billion passenger-kilometres, and in the 1990s this growth stabilized. Across the whole period it has been the growth in car travel, which has explained almost all the increase. Travel by bus and coach and the pedal cycle has declined by about 45%, and as a proportion of all distance travelled these three modes now only account for a mere 6%. Rail travel has remained remarkably stable over this period, with some signs of recent increases, and there has been a rapid growth in air travel.

Table 1.1 Increases in travel in Britain, 1960–2000 (billion passenger-kilometres)

Mode	1960		1970		1980		1990		2000	
Bus and coach	79	27%	60	15%	52	10%	46	7%	45	6%
Cars, vans, taxis and motorcycles	150	53%	301	75%	396	81%	593	86%	626	86%
Pedal cycle	12	4%	4	1%	5	1%	5	1%	4	1%
Road total	241	85%	365	91%	453	92%	644	94%	675	92%
Rail	40	14%	36	9%	35	7%	41	6%	46	6%
Air	0.8	0.3%	2	0.5%	3	0.6%	5	0.7%	7	1%
Total	282		403		491		690		728	
Average distance travelled per person per year	5,640		7,500		8,000		10,400		10,900	

Source: Department of Transport (1992a) and DETR (2000a)

This increase in mobility reflects the increased motorization and the changes in location of facilities, which have become more dispersed and distances between them have increased. These longer journeys make it less convenient to walk or cycle, or to use the bus. Particularly over the last 20 years, it has become increasingly difficult for planning authorities to refuse planning permission on the basis that access is inadequate. Many authorities have attracted development through the promise of free parking on sites that are often only accessible by car. There seems to be a basic inconsistency between a strategy for location decisions that encourage greater use of the car and longer journeys, and one that provides local accessibility and is compatible with the need to reduce the use of resources. Increasingly transport plays a central and crucial role in determining who gains access to what opportunities, and as such forms an important part of every person's quality of life.

The explanation for this phenomenal growth in travel, and in particular the increasing dominance of the car, is complex. Part of the reason is the decentralization of jobs and other activities from the city centre as a result of the changing industrial base of the country and because of the high land prices in city centres. Population has tended to follow this movement with extensive suburbanization and, more recently, the desire to live out of the city altogether. Cities have become less dense and more spread, and the car has facilitated all of these changes. The second major explanation is the growth in the economy and the real increases in income levels over the last 30 years. There is a strong correlation between Gross Domestic Product (GDP) and distance travelled by all modes of transport and a weaker relationship between GDP and car ownership (Figure 1.1).

Both personal travel and the transport of goods and freight have almost doubled since 1970, a rise closely linked with economic growth. Car mileage figures are also modified by the real price of fuel. Since 1960, the real cost of fuel has remained constant, but it did reach peak levels in 1981–83 when it was some 25% more expensive than today. The reductions in the costs of motoring, combined with increases in real incomes during the 1980s have led to this growth in road traffic. Households spend 70% more in real terms on

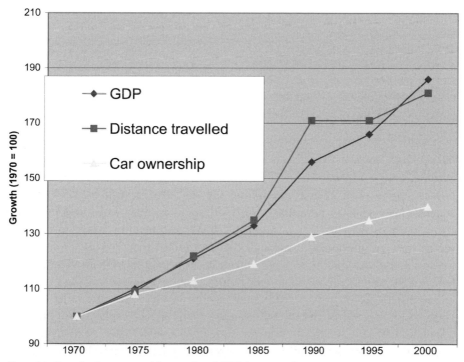

Figure1.1 Growth of car ownership, travel and GDP (1970–2000)

transport than they did 30 years ago, even though transport costs have risen more slowly than disposable income (DETR, 2000*b*). From the latest forecasts of traffic growth (DETR, 2000*b*), vehicle-kilometres will grow by 22% to 2010, with congestion[1] across the network as a whole expected to increase by 15% and by 28% on the inter-urban trunk road network.

Coupled with the spatial dispersion and economic growth arguments has been the changing demographic structure of the population. Britain's population is ageing, and today's elderly are the first generation of retired people, which has experienced mass car ownership and so can be expected to continue to be car oriented. With the tendency towards earlier retirement, a significant new group is emerging within the population, namely the 'third age of personal fulfilment' (Laslett, 1990). This group (aged 60–80) has ended the complex responsibilities of earning a living and raising a family, they are reasonably affluent, and so have money and time to spend on personal fulfilment. Many of these activities involve travel and it is here that one major growth area in travel can be expected in the next decade (Table 1.2).

Table 1.2 Summary of the main population trends, 1980–2000

Group	Laslett's Ages	Trend	Comment
School leavers and young adults	1st Age 0–20	Growth of 10% in the longer term	Decline has already taken place in 1980s and 1990s
Working age adults	2nd Age 20–60	Small increase of about 2%	Decline in 20–40 year age group in the 1990s, but growth in the 40–60 year age group
Age of personal fulfilment	3rd Age 60–80	Stable over the next decade	Growth expected in this group in the longer term
Full retirement and dependency	4th Age 80+	Growth of 25%	Major growth in next 40 years

It is only when people reach the age of 80 that full retirement and dependency begins to take place. This group of people, which accounts for about 23% of the elderly, will require special facilities and transport services which can accommodate their particular requirements, for example to be wheel-chair accessible or to have a person to accompany them. This group will not be able to drive and so will require public transport services or taxis or chauffeur driven private cars such as the service currently provided by many voluntary sector organizations. One possible development might be to design a vehicle specifically for the elderly to give them some degree of independence, perhaps similar to the battery operated tricycles, which are already available, or low speed electric cars with voice activated functions to ease the physical requirements of driving. Special routes could be provided for these low performance vehicles, which could be used by the elderly and perhaps children. With the age of consent being reduced (it is now 12 in the Netherlands), the age at which young people can drive a low performance vehicle may also be reduced, particularly if there is a novice period for newly qualified drivers such as operates in Australia. Such a change would allow greater independence for young people and reduce the need for parents to become chauffeurs with complex escort duties.

In addition to greater motorbility for young people and the growth in the elderly population, the age of retirement is also being reduced to 55 years or less. This means that apart from the natural growth in the elderly population, the number of people in the retired age group will increase through early retirement, and Laslett's third age would now cover the age range 55–80, not 60–80.

There are also changes in family structure with an increase in divorce rates, single-parent families and couples living together without children. These factors, together with the fall in fertility rates and the increase in the proportion of elderly in the population, have led to smaller household sizes. It is estimated that by 2016, about 8.6 million households in the UK will be one-person households (36% of all households). These demographic trends have an impact on the housing market with an increase in the demand for small housing units which may be located either in the city centres through the subdivision of existing larger properties or in the suburbs in purpose-built units. In each case it is likely that the ratio of car parking spaces to homes will be near unity, but if location is in the suburbs then the number of trips generated and the length of trips will both increase substantially. Saturation levels in car ownership will mean that every able-bodied person of driving age will have a car. The actual figure (16–74 years old) is likely to be about 73% of the total population by 2025, and some 90% of these will have cars – 650 cars per 1000 population. If special cars are produced for the young and the elderly, then this number might increase to nearly 800 cars per 1000 population, about double the present level of car ownership in most European countries.

Current patterns and trends

The dynamics of change in population mean that the spatial distribution of traffic growth is not uniform. It seems that the link, argued in classical location theory, between home and workplace has been broken as many people now move home for reasons not directly associated with work (Champion et al., 1987). For example, access to a job is not a consideration for retired people; there has also been a growth in the number of people working from home and in the proportion of households with two or more people working in different places; and a growth in the numbers of self-employed and the attractiveness of long-distance commuting.

The pattern of work journeys is no longer fixed by time of day or by day of the week. Destinations may vary as individuals visit the head office once a week, regional offices on other days, or are involved with site visits or overseas trips. Time for travelling may also vary. Regular patterns may be established, but not on a daily basis. Commuting patterns have become more complex with cross commuting becoming more important than commuting to city centres. For example, it seems that households now establish a residential base and career needs are met by commuting (Boddy and Thrift, 1990). Within a tight labour market (such as that which has existed in the south of England), there is often more than one person in each household who is employed, and complex travel patterns emerge as the transport system has to accommodate to this change. When interest rates are high and there is limited movement in the housing market, it is again the transport system that has to adapt, as people cannot move home and so develop long-distance commuting patterns. These patterns of change have been compounded by recent increases in house prices, as more low-paid workers (particularly those in the public sector) have been excluded from certain locations through price barriers. The conditions in the housing market and the increase in the number of economically active people, particularly of women, have led to more travel, longer travel distances and a new complexity in travel patterns.

There have also been fundamental changes in occupation and work status with the move away from semi- and unskilled jobs in manufacturing towards information and service based employment. The increased participation of women in the labour force is well known, and this trend has continued with about 90% of the one million increase in the labour force in Britain being women (1988–2000). Associated with this increased participation of women has been an increase in part-time working and to a lesser extent job sharing. Some of this growth may result from women rejoining the labour force after they have raised their family and from those mature women who have gained new skills through retraining. In Britain the growth in professional, technical and managerial occupations averaged about 1.7% (1988–2000), more than twice the average for the economy as a whole (0.7% per annum).

As these changes take place travel demand patterns also respond. Women often have considerably greater constraints on their activities as they have a multi-functional role involving getting the children to school, doing the shopping and carrying out other domestic activities. If employment is added to this list, the number and range of trips generated will also increase. If men take on more of the family responsibilities, more traffic will again be generated. Lifestyles are becoming more complicated for all members of the household not just the adults. Children too are evolving complex social patterns based around the school and home (Grieco *et al.*, 1989).

However, increased participation in the labour force has to be balanced against the shortening of the working week and increased levels of affluence. Economically active people now have to support a growing number of pensioners, as the labour force participation of people over 65 decreases and as the level of affluence increases. Again, these trends are likely to generate more travel both nationally and internationally. In the UK, 40% take no holidays at all (Government Statistical Services, 2000), but over 35% take two or more holidays (1998) and expenditure on leisure now accounts for over 20% of all household expenditure. Nearly all those in full-time employment have over four weeks of paid leave and a third have more than five weeks paid leave (1998). The number of overseas holidays taken by UK residents by air and by sea or through the Tunnel has risen from 31 million in 1990 to 54 million in 1998, and holidays within the UK now exceed 120 million. The package holiday and changes in air transport have revolutionized the way in which the world is viewed.

Instead of taking only one holiday each year, families and individuals are taking two or three holidays – a main summer holiday overseas, a winter holiday skiing and a third holiday based in their own country. In addition, days are taken off, often on Fridays or Mondays, to give short bridging breaks or long weekends. These trends are likely to continue and increase. It has been estimated (Masser, Sviden and Wegener, 1992) that by 2020 leisure activities may account for as much as 40% of all land transport (in terms of kilometres travelled) and 60% of air travel across all European countries.

The second trend is the increase in affluence and the importance of self-development and achievement. Apart from the activities of the mobile early retired groups, there are many other groups involved in a wide range of activities either of a social or voluntary nature or of a challenge (for example sporting achievement), or of an environmental nature. In each case, time, skill and knowledge is given to the activity and no payment is received. One reason for this self-development has been the growth in real incomes over the last 10 years and the increased levels of inherited wealth. As a result of the unprecedented increases in levels of house prices over the last 30 years, home owners can now spend that capital gain or pass it on to their children. Alternatively, when individuals trade down in house size or move to cheaper areas, capital is released so that more consumer spending can take place. Similarly, people are borrowing more against the actual or expected rises in house prices. Wealth, together with income and available credit, fuelled the increase in consumer spending which was a feature of the 1980s and 1990s. Increased rates of growth in car ownership, including a large increase in the provision of company cars, and increased levels of participation in a wider range of activities have all contributed to the growth in the number of trips, the range of destinations and the distances travelled.

These trends were curtailed through recession in the mid-1990s, and this resulted in falling house prices in some locations, leaving homeowners with negative equity.[2] This situation again makes it difficult to move home and with high levels of unemployment the housing market stagnates. The impact on the demand for transport has been a slowing down of the expected growth in traffic, and the development of new travel patterns resulting from people searching for jobs and moving less frequently. Once the gearing between house prices and wages is readjusted, the underlying increase in values continues, but at a lower level. The rapid increases in house prices experienced during the 1980s are not likely to be repeated, and the readjustment process was completed in the mid-1990s and prices are now again on an upward curve.

The third change in lifestyles has been the use of the home as an office or workplace. Much has been written on working from home together with the dreams (or nightmare) of the electronic cottage (Moss, 1987; Nilles, 1988; Miles, 1989). It has been estimated that 20% of all urban trips, including about 50% of skilled workers' trips, could be by telecommute, but that only 5–10% of these potential telecommuters would actually make the change (Button, 1991). The reality is somewhat different, and the best estimates suggested that home-based teleworking reduced total UK miles by some 1% (1993), with a long-term potential for some 5–12% of total car use to be substituted by telecommuting (Gillespie *et al.*, 1995). Monitoring at Surrey County Council's Epsom 'telecentre' found that the length of commute journey was reduced by 19% and the time taken by 36%, which reflects the levels of congestion currently experienced (NERA, 1997).

The hard evidence of such a change in travel patterns does not seem to be present in Britain or elsewhere. There has been very little change observed in the United States (Hall and Markusan, 1985), although empirical evidence from California does seem to support the argument for modifications to travel patterns (Pendyala *et al.*, 1991). In the California telecommuting project it was found that travel was substantially reduced. On telecommute days, the telecommuters made virtually no commute trips, reduced peak period trip-making by 60%, vehicle-miles travelled by 80%, and freeway use by 40%. As important was the finding that the non-work destinations selected were closer to home, not only on telecommuting days but on commuting days as well. The Californian experiment provides an important indication of the fundamental changes in travel patterns that might be obtained from telecommuting for one or two days per week. This change would reduce commuting and perhaps disproportionately long-distance commuting as this is where the advantages of telecommuting are most apparent. However, it is also likely that the long-distance commuting journey would be replaced by a series of short-distance trips to alternative locations for different purposes. Overall, travel distance may be reduced, but not the number of trips. In Scandanavia and the USA, it was found that working in 'telecentres' or local offices resulted in commute miles saved each time a worker used the centre ranged between 38.5 and 150 miles (62 and 240 km), or about 93 miles (149 km) on average (Bagley *et al.*, 1994).

These changes in lifestyle identified here will not be felt by all society equally. Some will not be directly impacted at all, whilst others will be, or have been, impacted by each change. It would seem that, as with all social change, it is the affluent that will be impacted first. Those on fixed or low incomes, and those who do not have the knowledge to react to technological change may only benefit in the longer term. The net result of lifestyle changes may be an increased polarization between different groups within society. On the one hand, there will be those affluent individuals with increasing leisure time who will be technologically literate – these information rich people will have increased mobility. On the other hand, there will be poor individuals on low or fixed incomes or unemployed, with no resources or leisure time who will not be technologically literate – these information poor people may possibly have reduced levels of mobility. The distributional impacts of social change may become even more significant in post-industrial society.

International comparisons

In a general sense, the pattern of change found in Britain is reflected across the other 14 European Union countries (EU) and those in the European Economic Area (EU countries and EFTA). All countries are moving to the post-industrial era, with increased leisure time, greater affluence and the extensive use of technology. Estimates suggest that the population in the EU would stabilize at 375 million in the year 2000, but this figure does not, however, take account of international migrations. It is expected that the current low fertility rates will be maintained. The most significant growth will take place in the elderly population, again as a result of the increase in life expectancy and through the tendency to retire earlier. The number of elderly in Western Europe (over 65 years) will rise from 13% (1985) to 20% (2020), and for all Organisation for Economic Cooperation and Development (OECD) countries the number will increase over the same period from 85 million to 147 million. As in Britain, average household size is expected to continue to fall, due in part to reductions in the fertility rate, but also as a result of increasing divorce rates and births outside marriage. In the 1970s divorce rates doubled in Belgium and France, and trebled in the Netherlands. By 1986 births outside marriage accounted for nearly half the total births in Denmark and Sweden (Masser, Sviden and Wegener, 1992).

The international movement of labour is still considerable between all

European countries and from East to West as well as from South to North. The increased participation of women in the labour force is also apparent in European countries with a similar increase in part-time working. This increase has been particularly marked in the Mediterranean countries (for example, Italy), but the apparent national differences may be due to different attitudes and traditions which are likely to remain.

When patterns of mobility and movement are examined across the EU member countries, similar variation is apparent. Over the 10 years 1988–1998, there was an increase of 30% in the numbers of cars and taxis in the fifteen EU member countries, with car ownership averaging 450 per 1000 population. Travel measured in vehicle-kilometres and passenger-kilometres has also increased by a similar amount with the total amount of passenger travel per head of population increasing to about 12,000 kilometres per annum (Table 1.3 and Department of the Environment, Transport and the Regions, 2000a).

However, internationally the overall pattern has considerable variation and this variation cannot be explained by population, area, density or economic factors such as Gross Domestic Product (GDP) or purchasing power alone. The greatest increase in car ownership over the last 10 years has been in the peripheral EC countries with low levels of GDP per capita (Portugal, Ireland, Greece and Spain). Growth has also been high in Germany (as a result of unification and equalization), Austria and Japan, but mainly from a higher base. Sweden has had the lowest levels of increase, followed by Denmark, Finland, the Netherlands and Norway. The other countries, including the United Kingdom, France, Italy and Belgium have below average growth (under 30%). Even in the United States the growth in vehicle ownership has been nearly 20% and the assumed saturation level of 650 vehicles per 1000 population has now been reached (Table 1.3 and Department of the Environment, Transport and the Regions, 2000a).

Over the next 20 years there will be a further significant increase in car drivers and in the number of cars (averaging between 60% and 80%) in many European countries, to about 550 cars per 1000 population or similar to the level in the United States in the 1980s (Banister, 1992c). This will mean an increase of about 40 million cars in the fifteen EU countries (+24%), and much of this growth will take place in households where there is already one car. As in the UK, it is unlikely that the road capacity will increase by anywhere the same amount and so congestion will increase.

Growth has also taken place in car use with Spain and Portugal both recording increases well above 50% (the EU average was 35%). All other EU countries had a growth in car use very close to the EU average, except Austria, the Netherlands and the Scandinavian countries (Table 1.3). The use made of each car has also increased in many countries, with Denmark, Italy, Portugal

Table 1.3 International comparisons, 1988–1998

Country	Car and taxi ownership per 1000 population 1998	Growth in car ownership (%) 1988–1998	Growth in car traffic (%) 1988–1998	Average km per car per annum 1998	Percentage of travel distance by car 1998
Austria	481	40	10	11,000	76.6
Belgium	441	25	37	12,200	83.4
Denmark	343	10	18	20,300	77.8
Finland	385	11	16	19,000	82.6
France	456	19	26	14,300	85.2
Germany	508	44	36	12,600	84.5
Greece	255	85	38	–	74.7
Ireland	311	53	37	25,000	80.0
Italy	545	27	37	14,900	82.3
Luxembourg	610	45	33	13,500	87.7
Netherlands	376	12	6	15,300	83.7
Portugal	320	52	73	14,800	85.9
Spain	409	49	126	9,300	84.4
Sweden	426	9	13	15,000	85.1
United Kingdom	404	20	27	15,700	88.7
EU15	451	30	35	13,900	84.4
Norway	407	10	16	14,700	87.2
Switzerland	476	22	23	13,300	80.8
Japan	396	62	32	9,500	63.1
United States	574	20	31	15,600	96.0

Source: Department of the Environment, Transport and the Regions, (2000a)

and Spain all recording more than 10% growth in the annual kilometres travelled per vehicle. The only two countries recording any real reduction in the use of each car are Norway (–4%) and Luxembourg (–7%). In all developed countries the position of the car is dominant with over three-quarters of passenger-kilometres being made by car, and the EU average is nearly 85%. The only exception is Japan (63%) where over 30% of travel is by rail, but even here there has been a substantial growth in car use (+80%) over the period 1988–1998. These major changes in travel patterns have not just been happening over the last 10 years, but since the mid-1970s in Europe and earlier in the United States. They illustrate the levels of car dependence in all developed countries and the scale of any change in policy required to shift modal split back towards public transport and the green modes of transport (that is, cycling and walking).

The importance of the transport sector in the EU cannot be underestimated – it accounts for 7% of GDP, 7% of jobs, 40% of public investment, and over 30% of energy consumption. These figures only record its direct contribution, and do not take account of transport induced activities nor its role in the overall functioning of a modern society (Group Transport 2000 Plus, 1990).

   Growth is also apparent in freight across the EU and other developed countries, but the rate of growth is lower than that of the car. In the ECMT-West (this includes the 15 EU countries plus Switzerland, Norway and Turkey), the total amount of freight tonne-kilometres moved has increased by 30% (1990–1998). Most of this growth took place in road freight (+38% in the ECMT-West), with little change in rail, inland waterways and pipelines freight. From an environmental and energy perspective these trends are disconcerting as the growth has taken place mainly in environmentally damaging road transport which is also energy intensive. The greatest growth rates over the decade in road freight have been found in Germany and Spain (>50%), and in Austria, Belgium and France (>40%). The increasing dominance of road freight is expected to continue over the next 10 years (to 2010) with a 74% increase in tonne-kilometres carried by roads out of a total increase of 51% (Group Transport 2000 Plus, 1990). These levels may be underestimates if the Single European Market succeeds in eliminating the barriers to movement and in harmonizing taxes between EU Member States, and if the expansion of the EU takes place during this period.

   Modern industry can now locate almost anywhere as it is not dependent on a single source of raw material inputs. Similarly, markets are national and international, not local. One of the results of this location flexibility has been an increase in travel demand. Investment in new infrastructure has improved productivity, reduced transport costs and strengthened the attractiveness of particular locations as there still seem to be substantial agglomeration economies (Banister and Berechman, 2000). The European high-speed rail network and the Channel Tunnel are two examples of major infrastructure investments which have opened up new locations for industrial development. Even in Southern European countries where agriculture is still a relatively large source of employment, the growth in new markets and greater transport accessibility have changed production methods, and increased efficiency and specialization. Companies have become transnational in order to exploit local labour cost differentials between individual countries and to compete in world markets (Roos and Altshuler, 1984).

   Technological change has reduced the effects of physical distance and allowed further decentralization of lower-order back office functions where cheaper labour can be used, and it is only the front office functions that need to be located in the city centre with a highly skilled and expensive labour force (Goddard, 1989). Many inventory, financial and communication transactions can now be carried out remotely. However, the impacts of technology are not equal across all sectors as they relate to the functions of individual organizations and the quality of their links with high quality and expensive computer infrastructure networks.

   High-level knowledge-based activities and skill-intensive tasks may be concentrated in a few core cities and regions, while low-skilled standardized production tasks are carried out in the peripheral areas within individual countries and also within the peripheral areas of Europe as a whole (Masser, Sviden and Wegener, 1992). The level of interaction between organizations both nationally and internationally, and within spatially separated parts of the same organization, has increased dramatically by all means of communication, including transport. The greatest unknown at present is the impact of Eastern Europe markets both as opportunities for increased sales and as locations for peripheral factory locations to exploit cheaper labour costs. Growth and affluence may be concentrated in particular countries or in particular areas such as 'the banana' from South East England through Benelux, South West Germany and Switzerland to Northern Italy, or 'the sunbelt' along the northern part of the Mediterranean. The opening up of Eastern Europe will certainly move the centre of gravity of Western Europe to the east.

The uncertainty created by the move to post-industrial organization and the structural changes taking place in industry, together with technological innovation, make it difficult to identify the actual impact of these factors on transport demand. In addition to the economic and technological revolutions, there have been unprecedented changes in the political boundaries with the opening up of Eastern Europe and the Single European Market.

Congestion was probably the most talked about transport issue of the 1990s with 'Gridlock' occurring regularly in city centres, suburban centres and more irregularly in rural areas. Experience in the United States may be informative as car ownership levels are high (580 cars per 1000 people), and all forms of gridlock are common and the demographic factors are already being felt. The official view (Federal Highway Administration, 1988) seems to be that the key determinants of future demand are population, age and gender, percentage of driving age population with a licence and personal income. At present nearly 90% of the adult population have driving licences and there are, on average, nearly two vehicles per household. Distance travelled by residents averages at over 29,000 kilometres per annum. These levels are much higher than those in the European Conference of Ministers of Transport (ECMT) countries, and the consequences of the same trends occurring in Europe may be severe as population densities are much higher and the infrastructure is less well developed (Banister, 1999).

However, not all the evidence is negative. In a most interesting paper, Lave (1990) argues that trend-based analysis is an inappropriate generalization from a highly atypical period of history. As vehicle ownership in the US is reaching saturation and nearly all the driving age population will have vehicle access, the growth rate of vehicle use will decline. He also suggests that the growth rate in vehicle travel will be much lower. While most analysts have been concerned with the consequences of demographic change, they have missed the structural shift in the demographics of car ownership and car use. This shift has led to a disproportionate growth in the vehicle population, but this stage has now ended as the demand for cars is saturated.

However, this optimism from the United States may not be appropriate in Europe, as car ownership has not yet reached the levels found in the US (Table 1.3), and even in the most affluent countries car ownership only reaches between 50 and 70% of the assumed saturation levels of 650 cars per 1000 people. Secondly, there is a link drawn by Lave (1990) between car ownership and car use, which he claims is a stable relationship. If there is no further increase in car ownership, he argues that there will be no increase in the use of those cars. Evidence from Britain and other European countries suggests that trip lengths have significantly increased, and that the lower mileage recorded by second cars in car-owning households is more than outweighed by the increased mileage recorded by households obtaining their first car. In countries where car ownership is still increasing and where structural changes in the economy are taking place, both the numbers of trips made and the distances travelled will continue to increase, leading to greater congestion.

Within the overall pattern it seems that certain sections of the population may travel more by car. Women and the elderly are two groups which have traditionally driven less than other people. The changes in women's participation rates in the labour force, greater independence and the increase in 'non-standard' households would all suggest that their patterns of travel would become more similar to those of their male counterparts. Similarly, with the increase in life expectancy, health, aspirations and affluence of the elderly, one would expect that they would both keep the car as long as possible and make greater use of it in their extended retirement.

These arguments, at least from a European perspective, would suggest that the complexity of demand for travel will continue to increase. For example, due to the changes taking place, it is unrealistic to expect that elderly people in the future will have the same travel patterns by mode as a similar elderly group today. There are at least three types of demographic change which would support the argument that trip rates by mode for particular groups will not remain stable in time.

The first factor is that present day expectations and travel patterns will influence aspirations in the future. This cohort effect will be most apparent with the elderly who are the first generation that has experienced mass car ownership and so can be expected to continue to use that mode as long as possible. The second factor relates to changes in lifestyle, the growth in leisure time, the high value now being placed on the quality of life, and the importance of stage in lifecycle. Lifecycle changes refer not only to the four basic conventional groups (married couples without children, families with young children, families of adults, and the retired – Goodwin, 1990), but to the wide range of unconventional groups (for example, single-parent families). Changes in life style and lifecycle effects have had fundamental impacts on the range of activities that people require, the increasing complexity of travel patterns, and the increase in travel distances. Complementary changes have also taken place with the structural changes in the economy and changes in the distribution of industry, commerce and retailing which have tended to follow the decentralization of population. The final factor has been the increase in levels of affluence and the unprecedented growth in car ownership levels. Some of this affluence has resulted from the growth in Western economies, but the greater part has been the growth in savings and wealth from property value increases. That new wealth is likely to be used by the newly retired elderly or passed on to their next generation. In Britain, it has been estimated by the Household Mortgage Corporation that inherited wealth from the sale of property was £8 billion (1990) and that this has increased to £29 billion in 2000 (Hamnett *et al.*, 1991 and Hamnett, 1999). It is unclear what proportion of this money will be invested rather than spent on consumer products such as cars or on other activities such as leisure which involve travel.

In Britain nearly 70% of homes are owner occupied, and this contrasts with European levels of owner occupation of between 30% and 40%. There has been a growth in owner occupation across many European countries, including the Netherlands, Sweden, Switzerland and West Germany, during the last 20 years. Although there has not been evidence of major increases in European house prices on the same scale as those in the UK, this may be changing as prices in many EU countries are now rising at 5–10% per annum. With lower levels of house ownership and lower increases in house prices (until recently), the private capital tied up in European housing may be less than that in Britain and so the levels of inherited wealth will be less.

For all these reasons the prediction of travel demand is difficult, but it is clear that demographic changes should form part of the analysis along with historical trends and changes in the economy. This suggests that travel demand will continue to increase at about 2% per annum in the wealthier EU countries, and at 3% per annum in the poorer EU countries. The assumption underlying this discussion is that there is a continued desire to travel, but there may also be a limit to that desire. With the increased levels of congestion on all transport modes, and delays at termini and interchanges, people's appetite for travel may decline, particularly where they have a choice. Quality of life factors become more important with increased affluence and leisure time, and travel may not provide an attractive choice. However, this limit may only be apparent with particular groups in affluent post-industrial Western economies. It is likely that any reduction in one group's appetite for travel will be more than compensated for by another group's increased propensity to travel.

It seems that the demand for travel will continue to increase but that the nature of that demand may change as a result of increasing wealth, demographic change, industrial restructuring, technological innovation, and the growth in leisure activities. The problem here is in unravelling the complexity of these issues so that the effects of one group of factors can be isolated from all the others. Similarly, there is a range of policy instruments which can be used to influence levels of demand and mediate between the different interests. Two main conclusions arise from this introductory review. Firstly, the overall levels of demand will continue to increase and the private car will accommodate most of that growth. Secondly, the composition of the demand will be significantly different as society becomes less dependent on work related travel and more dependent on leisure travel, and as groups within the population which have traditionally been seen as having low levels of mobility now start to have much higher levels of mobility.

The basic policy question then becomes whether and how that increase in demand can be accommodated given the economic, social and environmental costs that will be incurred in developed nations if these mobility trends are allowed to develop unchecked. The alternative must be some form of planning in the allocation of resources and priority to the more efficient modes. The implications for transport operators are considerable if both the demand for public transport is uncertain and the underlying stability of traditional public transport markets is being questioned.

Similarly, there are important decisions for policy-makers in deciding whether to increase the capacity of the road system through new investment or to manage the existing infrastructure through pricing and controls on the use of the car. Action is required, as no action will result in increased congestion and inefficiency in the transport system. Growth in demand takes place continuously, yet growth in the capacity of the transport system is discrete and often takes considerable time for implementation.

The conclusion reached here would strongly suggest that there must be some overall strategic view that links changes in transport demand to changes in demographic factors, the economy, technology, and in land-use patterns on the one hand, with a concern over the environmental and quality of life factors on the other hand. A single action perspective along only one dimension reveals only part of the picture. To understand and to respond to the whole picture, one must investigate the interactions between all relevant sectors to produce a composite strategic view.

Conclusions

Society is in transition from one based on manufacturing to one based on services, leisure, self-fulfilment and travel. Travel is playing an increasingly important role in our lives with the development of a highly mobile car-oriented lifestyle. Transport planning has also had an important role in understanding the processes of change, in anticipating future growth and in evaluating options. The methods of analysis have evolved from aggregate models applied on a city-wide basis to disaggregate approaches developed to investigate particular problems. In parallel with the development of analysis methods have been the changes in the statutory planning process within which analysis must be placed. Over the last 40 years this planning process has undergone a radical change, with the underlying philosophy switching from notions of welfare, planning and the availability of transport for all people to one based on notions of the market, competition and the payment of the full costs of transport by the traveller.

The methods and approaches for the analysis of transport developed in the 1960s and the 1970s were appropriate to large-scale systematic investigation where the responsibilities of central and local government were unambiguous. There was a clear statutory planning framework which was comprehensive in its scope, and models could play a central role in the decision process by testing alternatives and evaluating options. Large data sets were available and standard procedures were developed, with transport planning being seen as a technical process. The concern was over the production of proposals supported by comprehensive analysis.

With the reorganization of local government in the 1970s and the widening of expertise, some of the basic assumptions in large-scale analysis methods were questioned. The public wanted transparency in analysis with methods being comprehensible to them and commanding their respect. The notion of the expert and the technocratic nature of the transport planning process were questioned, as were some of the basic assumptions on growth and the absence of what were acknowledged to be key variables (for example employment distribution and land values). This uncertainty coincided with reductions in public funds available for transport investment, and with the restructuring of employment and inner-city decline. Transport planning was no longer primarily concerned with the production of major road schemes, but with evaluation, value for money and the monitoring of change. Coupled with practical concerns over the usefulness of the comprehensive methods developed in the 1960s and the 1970s, and changed political priorities, was a fundamental theoretical impasse. It was argued that systems analysis (including transport planning analysis) was functionalism carried to an extreme and highly complex form (Sayer, 1976). All that had been achieved was the maintenance of the existing social system, and radical alternatives or even the gradual switch to a post-industrial economic base could not be encompassed in such a mechanistic form of analysis.

Transport planning was not prepared for the radical policies of the 1980s with the weakening of links between transport and land-use, and the emphasis on finance-driven economic policies. The 1980s saw the demolition of the statutory planning process, the abolition of the Metropolitan County Councils (and the Greater London Council), and the fragmentation of responsibilities in local government. This resulted in an increased central-ization of power with central government having direct control over local budgets. To achieve this magnitude of change there has been a massive programme of legislative reform both in planning and in transport, with the ensuing uncertainty and disruption. The primary aim of the changes has been to reduce the levels of public expenditure in transport, particularly on revenue support. Where expenditure was available, it should be targeted on capital projects with contributions being sought from the private sector.

From 1965 to 1990, there were three major organizational changes affecting local government (1968, 1974 and 1986). If it is assumed that preparation and adoption take two years each, then 12 out of the last 25 years have been spent preparing for, implementing and adapting to change. If minor changes are also included (1980, 1983 and 1988), then 15 years are accounted for (Tyson, 1991). Since the change of government in 1997, there has again been a period of uncertainty as a new tier of regional government has been introduced in England, together with the devolution of power to the Scottish, Welsh and Irish assemblies. More powers and resources have been distributed to the regions, and cities are again being given greater responsibilities through the election of Mayors and local assemblies (for example, the Greater London Assembly in London).

The problems facing transport planners are considerable. In the 1960s and the 1970s there was a clear link between policy, planning and analysis, but the 1980s saw a switch to policy-led decisions based more on ideological concerns than analysis. Physical planning as reflected in the land-use and transport planning process is most relevant in a stable environment with public owner-ship (and a statutory monopoly) of the main transport operators, extensive government intervention, control of prices, and of the levels and quality of service. This approach had its origins in the statutory planning system and the nationalization programme set up after the Second World War as part of Herbert Morrison's plans for the reconstruction of British industry. There were clear welfare objectives in economic policy with the state playing a key role in the reconstruction process. Decision-making was *pluralist* with power being widely distributed between groups without any one group having overall control, and the state's role was essentially neutral as it acted as a mediator between the different interest groups.

Corporatism developed in the 1970s as a more positivist approach to policy-making as well as a response to world recession, high inflation and unemployment triggered off by the rise in oil prices in 1973 and 1979. This period was one of retrenchment where the state formed a series of reciprocal relationships with major organized interests. It seemed at the time to be in the interests of both national and business concerns, which were seeking state protection against competition. This was a defensive mechanism in volatile world markets. Industry operating as monopolies or quasi-monopolies within the domestic sector used the state to plan for growth with minimal competition within either domestic or international markets.

State involvement as partners with industry or as the controlling influence in nationalized enterprises now stood at the centre of decision-making. However, this position was inevitably short lived, as a result of both external and internal events. The 1970s were uncertain times in economic terms as trade stagnated, demand fell, and unemployment as well as commodity prices rose. The response from industry was to cut back on output, reduce manpower levels and levels of investment. This in turn led to labour disputes and demands for higher wage settlements to match increased prices. At the governmental level taxation was increased, the balance of payments deficit was increased, and the standards of living were at best maintained.

It was against this background that a radical alternative was introduced, namely the move to the *company state* and eventually to the *contract state* where the role of the state is to facilitate the operations of private companies. Initially, the aim was to go towards a market economy based on the well tried neoclassical economic principles. The role of government, both central and local, was to be reduced to that of a facilitator so that the nationalized industries could be returned to the private sector. It was argued that organizations are more efficient in the private sector with the normal commercial pressures that competition brings. Coupled with the privatization programme was a parallel programme of regulatory reform, reductions in trade union powers and the abolition of organizational structures which seemed unnecessarily bureaucratic. The role of the company state is to ensure that the conditions of competition are maintained and that public concerns over such issues as safety and the environment are met. The second stage involves the dismantling of government itself as the role of the state is reduced with the market becoming the arbiter and not the state. This is the contract state where agencies are set up to ensure that competition is fair and that consumer interests are balanced with those of the newly privatized companies. This aspect is particularly important where state monopolies have been replaced by private monopolies.

Most recently, there seems to have been a return to the *corporatist* approach with government working in conjunction with business, both in running transport services and in financing new investment. But this new form of corporatism is very different to the defensive alliances forged in the 1970s. The 1990s version treats business as partners and as investors in the infrastructure, as resources from the public sector are limited. Taken to its limits, it suggests that the road infrastructure should no longer be a public good and provided 'free' at the point of use to users, but that all users should pay for access to the network. This has happened for many years for access to the public transport network, where fares have covered both the operating costs and made some contribution to capital costs. Access to the road may be charged for in addition to the operating costs of cars and contributions to exchequer revenue. However, this future has not been clearly formulated as a workable strategy, even though the Ten Year Plan (DETR, 2000*b*) goes some way to promoting the new relationship between government and business.

The basic argument presented in this book is that transport planning must respond to these new imperatives. The approach adopted in the 1960s and 1970s was appropriate within the stable political environment of that time, but even in the mid-1970s planning for growth was replaced by substantial concerns over whether the methods were equally appropriate for investigating the transport implications of industrial decline and restructuring. In the 1980s and 1990s society moved from manufacturing to service employment, from mass production to information technology, to the globalization of the world economy, with major demographic changes in the population and the increases in leisure time. The information revolution is likely to be as significant as the agricultural and industrial revolutions.

The book is divided into two halves, the first presenting a retrospective of the evolution of transport planning. The development of transport planning in Britain is outlined together with its rejection in the early 1970s and its subsequent re-emergence in basically the same form in the late 1970s. This review is complemented by a shorter section on the similar developments in planning analysis. Throughout this period there was an almost parallel evolutionary process. It was during this period that evaluation became increasingly important with growing limitations on public budgets, and the necessity to ensure that value for money was being achieved. The final part of the retrospective presents a view of contemporary transport planning where the market alternative to transport provision is presented, based on the premise that the state should be a facilitator and that intervention should be minimized. Clear conclusions are drawn for transport planning analysis which are intended to act as pointers for the early twenty-first century. These include the issues of infrastructure, congestion, forecasts, regulatory reform, and the environment, all of which suggest that some form of intervention is required.

The second half of the book is a comparison and prospective. The first part explores experience of transport planning in other countries, taking material from the United States, France, Germany and the Netherlands. It also looks at European approaches to evaluation in transport. The purpose of this comparison is to place the British experience in a wider context and to examine the reasons for different approaches to essentially similar transport issues. It examines the processes by which ideas and techniques have developed and explores whether there has been any cross fertilization between countries. The prospective part discusses some of the new approaches which have been developed in the last 10 years. Apart from discussing methodological issues, it also comments on the new problems and the role that transport planning can and should have in the new millennium. It sketches out some of the key issues likely to challenge transport planning in the next decade particularly when placed in the context of radical political change, and demographic, industrial and technological changes, and the breaking down of international barriers in Europe. The role for transport planning has changed very substantially over the past 40 years and the prospects for the next decades are exciting as new challenges, such as the market-state relationship, the funding of transport infrastructure, and the planning framework, will all have to be confronted and resolved.

Notes

1 Congestion is defined as the average delay experienced for each kilometre travelled compared to driving speeds typical when traffic is light. For an interesting discussion on the notions of congestion see CPRE (2001).
2 Negative equity occurs when houses are worth less than when they were purchased.

Chapter 2

The evolution of transport planning: the 1960s and 1970s

Introduction to the transport planning process

Transport planning has evolved over the last 40 years, but with no clear theoretical foundations. Everyone is aware of the problems created by the increased demand for transport and most effort has been directed at finding methods of analysis with a practical, usually quantitative, output. This has meant that analysis has been empirical and positivist in its approach.

Initial developments were concerned with aggregate analysis and the efficiency of overall movement of people and goods. It was in the 1960s that the transport planning process evolved as a systematic method for 'solving' the urban transport problem. The classic deductive approach was adopted with the future state of the system being synthesized from a series of laws, equations and models. The transport planning process was intended to be comprehensive with the collection, analysis and interpretation of relevant data concerned with existing conditions and historical growth. The aim was to establish goals and objectives, to synthesize the 'current patterns of movement' within the city, and to forecast future demand patterns either with trend based changes or with a range of investment options. These alternative packages would be evaluated against the 'do nothing' situation and the goals and objectives set at the beginning of the process.

The structure of the transport planning process followed the systems approach to analysis and marked the move towards an analytical approach rather than decisions being based on intuition and experience. The broad structure of the approach followed that of the Chicago Area Transportation Study (1960), one of the first classic aggregate studies (Figure 2.1). This is the basic structure which is still used, albeit with many modifications. Expected vehicle and passenger volumes in the main travel corridors were estimated and increases in road and public transport capacities were proposed to accommodate those expected increases over the following 20 years.

As itemized by Thomson (1974) the basic process can be summarized in eight stages

◆   *Problem definition*: what is the problem and what are the planning objectives?
◆   *Diagnosis*: how did the problem originate with views from different perspectives (for example engineering and economic)?

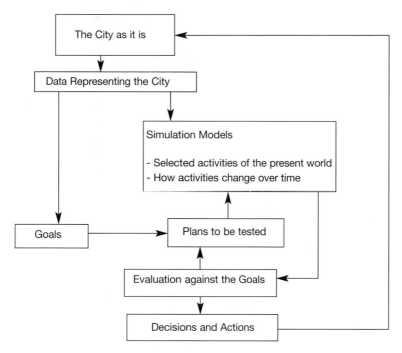

Figure 2.1 The structure of the Chicago Transportation Study (Source: Based on Creighton, 1970)

- *Projection*: forecast of what is likely to happen in the future. This is often the most difficult stage.
- *Constraints*: three main types of constraints limit the choice of alternatives (financial, political and environmental).
- *Options*: what are the range of options which can be used to achieve the planning objectives stated in the first stage?
- *Formulation of plans*: a set of different packages covering road and public transport alternatives.
- *Testing of alternatives*: usually through a modelling process to see whether each alternative can achieve the stated objectives and how each compares with other alternatives. Trip generation, trip distribution, modal split and traffic assignment studies.
- *Evaluation*: to assess the value for money usually through some form of cost benefit analysis or financial appraisal.

This structure took several years to evolve, but variations on the basic format have been applied to hundreds of transport studies that have been carried out all over the world (Bruton, 1985; Hutchinson, 1974). The Transport Planning Model (TPM) formed the central part of the transport planning process and was the testing of alternatives in Thomson's categorization. Conventionally the TPM is divided into four sequential, linked submodels:

- *Trip generation* is the number of trips associated with a zone or unit and consists of trips produced and trips attracted to that zone.
- *Trip distribution* is the allocation of trips between each pair of zones in the study area.
- *Modal split* determines the number of trips by each mode of transport between each pair of zones.
- *Trip assignment* allocates all trips by origin and destination zone to the actual road network. Separate allocations normally take place for each mode.

Other factors such as land-use and population changes are input exogenously to the TPM once it has been calibrated for the existing situation. It is sequential in that the output from one submodel is the input to the next. Information about transport networks, about the location of facilities and about the characteristics of households (for example car ownership and income) are all introduced into the model sequences at the appropriate stages. The output represents the transport system and this is used as the basis against which to evaluate alternative plans. The implicit conceptual foundations of the TPM are that the decisions made by each traveller follow this simple sequence – whether to make a trip, where to go, what mode to use, and which route to take. The model does not conform exactly to this decision process as the data are collected for zones and the TPM is carried out for the city as a whole. It is an aggregate model.

The development of the land–use transport study: the 1960s

Until the 1950s the method by which existing traffic levels was predicted was to extrapolate the growth to the forecast date. A simple growth factor approach was often used giving rise to unconstrained inductive planning. The first real conceptual breakthrough came with the work of Mitchell and Rapkin (1954) at the University of Pennsylvania. They put forward the theory that the demand for transport was a function of land use, and following on from this basic premise it was argued that if land use could be controlled, then the origins and destinations of journeys could be determined – 'urban traffic was a function of land use'. In the mid-1950s large scale land-use studies were carried out on this basis (for example the classic studies in Chicago and Detroit), with changes in land use being taken as the principal exogenous variable. In these large-scale aggregate city-wide studies transport planning was seen as a technical study based on what were considered to be rational principles. Plans could be produced on the basis of factual inputs and planners were able to move away from the 'value-strewn area of municipal politics' (Gakenheimer, 1976).

These early studies were developed to assess the changes resulting from new roads, and from new systems of traffic management and control which were being introduced. The early 1960s were marked by exponential growth in car ownership levels, growth in real incomes, cheap energy, and a vision of the expansive low-density city. It was also a period of considerable construction with extensive housing programmes and new town development. The optimism of this time was characterized by the view that the reason for congestion was a lack of road space and that this problem could be resolved by more road construction. The early generation plans were heavily road oriented with this naive belief that heavy investment in an upgraded road infrastructure would 'solve' the transport problem.

The first generation of traffic studies was carried out in Britain in the early 1960s. The basic approach has been to model the movement of people and goods in the study area, to identify where congestion is likely, and to predict the use that will be made of the transport network for a 'design' year say 20–25 years ahead, given no change in the supply and given certain specified changes. The alternatives were compared in economic and technical terms with a preferred system covering roads, public transport and parking policies being recommended for the area. These first generation studies involved cooperation between the local authorities, the transport operators and national government. The first and largest of these studies was carried out in London. The London Traffic Survey (1961–62) was intended only to produce road traffic forecasts for 1971 and 1981 and to relate those forecasts to the road networks envisaged for those years (London County Council, 1964). When the Greater London Council came into existence (1965) the narrow objectives of the study were broadened to cover a more comprehensive approach to transport analysis.

These urban traffic studies were complemented by the strategic road plan initiated after the Second World War and brought to fruition through a firm government commitment to a basic 1000 miles (1600 km) of motorways. The London-Birmingham (Ml) motorway was opened in 1959 with 72 miles (115 km) of high-quality dual carriageway road with grade separated junctions. Its economic benefits had been estimated in a seminal cost benefit study (Coburn, Beesley and Reynolds, 1960), and the political and public impetus for more roads had been created. As Starkie (1982) commented, the modified 'tearoom' plan of 1946 had become Conservative government policy (Table 2.3, p. 44). This period was characterized by the maxim 'predict and provide' with the corollary that cities had to adapt to the car. The shortfall in capacity meant that it was an engineering problem requiring some assistance from the economist on evaluation. The wider implications of road building seemed not to have been discussed and the whole process from problem formulation to prediction, evaluation and implementation was a technical exercise.

Certain events took place in the period 1963–1965 which in retrospect marked a fundamental change in both attitudes and approaches to the analysis of urban transport problems. The Buchanan Report on *Traffic in Towns* (Ministry of Transport, 1963) made two important contributions. One was the realization that large increases in road capacity tended to exacerbate the problems of traffic congestion and not solve them. Secondly, there was a realization that there were significant environmental disbenefits caused by traffic. Certain ideas were also put forward in the report. The principle of the road hierarchy and the notion of environmental areas formed the core of the Buchanan approach to the design of towns for the motor age. It is ironic that environmental capacity was interpreted by the Germans as instrumental in initiating ideas of traffic calming, but by the British as promoting an expanded programme of urban road building. The German equivalent of *Traffic in Towns* (Hollatz and Tamms, 1965) was less concerned about the environment but more concerned about the means to promote traffic flow through the construction of orbital and ring roads, and the need for heavy investment in public transport. Hass-Klau (1990) argues that Buchanan was ahead of his time and that it was only with the rise in the environmental movements of the 1970s that the Report's true impact was really felt. However, it could also be said that Buchanan was himself restating ideas, particularly on precincts and their similarity to his environmental areas, which had been put forward by Alker Tripp some 30 years earlier. Nevertheless, it was realized that environmental standards within any urban area automatically determine accessibility and that modification of this relationship would need considerable investment in the physical infrastructure to remould towns and cities for different levels of access (Ministry of Transport, 1963).

The second report received less publicity but was also pioneering in that it examined the possibilities of road pricing in urban areas (Ministry of Transport, 1964). The Smeed Report argued that people should be charged for the congestion they cause. Even though many of the economic arguments were accepted at the time, some are still open to debate now, namely the possibility of very low elasticities of demand, equity concerns and the impact that charging might have on the local economy. The possibility was rejected for technical reasons given the unreliability of metering systems and the costs of implementation. These technical problems have now been overcome, and as with Buchanan's ideas on environmental areas, Smeed's ideas on pricing are now being reconsidered. The problems of public acceptability still remain, but experiences in Singapore, Oslo, Trondheim and Bergen all suggest implementation is possible. Even London is now seriously considering cordon pricing (GLA, 2001). No decisions were taken on the road pricing issue and the government eventually produced a report on *Better Use of Town Roads* (Ministry of Transport, 1967) which effectively dismissed both road pricing and supplementary licensing in favour of tighter controls of parking and making better use of the existing road capacity through traffic management schemes.

However, there was now a clear view that traffic studies should not just be concerned with the construction of new roads, but that plans should include the use of the available capacity, public transport options and limitations on the unconstrained use of the car through control measures. This switch to transport studies and the balanced approach to cover all forms of transport was crystallized in the Circular 1/64 issued by the Ministry of Transport (1964). The Ministry would give technical advice and share the costs of these new transport studies. The key issue had become modal split and the relationship between the car and public transport, and the effect of different investment options on the balance between the two.

Even though these studies (for example the West Midlands Transport Study, 1964 (Freeman, Fox, Wilbur Smith & Associates, 1968); Merseyside Area Land Use Transportation Study – MALTS, 1966 (Traffic Research Corporation, 1969)) were broader based and included a wider range of land-use alternatives together with public transport networks, their investment proposals were still heavily in favour of roads. Moreover, the assumptions about growth in population and employment on which transport investment decisions were based seemed wildly optimistic. For example, the MALTS study team claimed that the report was based on 'a thorough review of how best to accommodate increases in population and employment over a period of time in which rising productivity and income are expected to lead to a substantial increase in the volume of goods on the road and in the number of cars used'. The MALTS suggested that population would grow between 15% and 28%, and employment by 15% (1966–1991). The main source of new employment would be in the service sector, and would be located in Liverpool city centre (Figure 2.2), regardless of land-use policies. Car ownership would rise from 12 cars per 100 people to 40, and 80% of households would have a car.

To accommodate this expected growth, five alternative land-use strategies were examined which combined one of three residential location strategies with either dispersed or centrally located employment. The study team opted for dispersed employment and proposed to accommodate population growth along radial corridors to the east of Liverpool. On the transport side, the MALTS team assumed that people would continue to be allowed to use their cars as and when they chose. Top priority was relief of central area congestion, and this led to the recommendation of a £287 million (1966 prices) roads programme together with a 50% increase in parking provision in the central area.

MALTS Methods and Assumptions

Methodology: Large study consisting of 7 separate surveys, 24 technical reports, with 5 alternative land use strategies examined along with 3 levels of car usage.

Techniques: Mathematical models developed for trip generation, modal split, trip-distribution and trip assignment, and applied to work trips, non-work trips, external trips and goods vehicle trips. Cost-benefit analysis applied to road construction over and above the (large) basic programme, the need for which was taken to be self-evident.

Assumptions

Population	% change 1961-91
MALTS area	35.2*
Liverpool	32.3*

Employment	% change 1965-91
Total	15.3
Manufacturing & wholesaling	14.5
Retailing	8.9
Offices & services	33.0
Other	– 5.9

These assumptions were derived from the most optimistic projections based on the 1961 census reports. The authors of the study thus assumed that growth of 7 per cent had already taken place between 1961 and 1966 (35 minus 28%). Source: Merseyside Area Land Use Transportation Study, Final Report, volume A, Traffic Research Corporation Ltd. 1969.

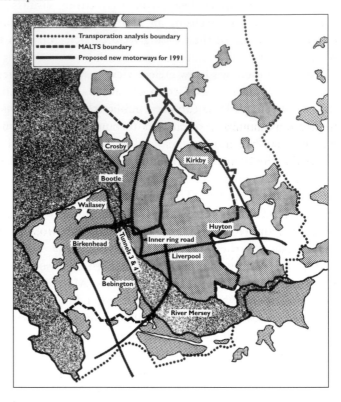

Figure 2.2 MALTS methods and assumptions (Source: Banister and Botham, 1985)

The road programme consisted of no less than 100 km of new motorway and 90 km of arterial roads. The team also recommended building a third Mersey tunnel (the second was then under construction) and thought that a fourth tunnel would be required in the 1990s. Three-quarters of all spending would concentrate on an area within 5 km of the city with an inner ring road forming the top priority.

Public transport improvements were also considered an essential part of the package with the local rail system being upgraded at a cost of £20 million. This involved the construction of an underground loop and link system, connecting the rail termini in Liverpool, together with the electrification of lines entering the city centre. As population was dispersed outwards, commuters were expected to switch from bus to rail; and the £20 million spent on the rail system was expected to reduce the need for spending on roads by some £10 million. By the late 1970s, 4000 park-and-ride places were assumed to be needed, and 40 trains per hour (in one direction) would be required under the river Mersey (Banister and Botham, 1985).

The MALTS study was typical of the early generation of transport studies in that a careful attempt was made to link transport with land-use planning. The traffic forecasts were based on the best available methodologies of the transport planning process, but there were few signs that the transport planners had been influenced by the land-use planners. Not one of the alternative land-use plans tested made any difference to the need for new transport infrastructure. The reason for this result was that significant city centre employment growth was assumed in all the options. Secondly, to maintain maximum flexibility in the plan, it was argued that the radial roads would allow the distribution of any further growth that might take place above the predicted levels. In retrospect two conclusions can be drawn. Firstly, the inclusion of several peripheral and radial routes in the plan could encourage the dispersal of population and employment, not the concentration assumed. Secondly, it seemed that the ambitious road plans were really a 'bid' for central government resources rather than a serious attempt to promote the best mix of transport and land-use policies.

The decade ended with a further shift in policy emphasis with the 1968 Transport Act. The concern here was not with a single best solution but with the allocation of resources and the best use of the existing infrastructure. Traffic management and parking controls were the two main measures used to increase flows and maintain speeds. Every authority over 50,000 population, and all traffic and parking authorities in the provincial conurbations were required to produce short-term traffic and transport plans to cover the period up to the mid-1970s (Circular 1/68 Traffic and Transport Plans and Table 2.1). Progress was slow and by early 1972 when they were

Table 2.1 Urban transport planning, 1968

The plans – need for measures designed specifically for
  (a) the relief of congestion
  (b) the help of public transport
  (c) building road safety measures into highway and traffic plans
  (d) the protection of the environment by traffic management

Urban authorities to prepare traffic and transport plans 'showing how they intend to relate their traffic and parking policies to their available road capacities and to their immediate and longer term policy objectives'.

These plans to be short term (to mid-l970s) and to take account of how other investment programmes (for example urban renewal, highway programmes, public transport, and parking) affect future traffic patterns.

The scope of the plans – broadly should state
  (a) the present and future situation (traffic, public transport etc) as far as possible in quantitative terms
  (b) the local authority's transport objectives against the town planning background
  (c) the criteria, guidelines and rules of thumb (however crude) which the local authority has applied in arriving at its plan
  (d) alternative strategies considered
  (e) the chosen plan and its cost

Source: Ministry of Transport (1968)

superseded, only 50–60 out of about 140–150 authorities had submitted plans. It was the plans for urban motorways and other large-scale road schemes that were capturing both attention and the headlines (Starkie, 1973).

Rejection of the 1960s approach

The optimism of the early transport studies was overturned in the early 1970s with a movement against transport planning and large-scale road building programmes. The process was no longer just seen as a technical exercise for the expert and his advice could no longer be accepted automatically. Other considerations apart from the technical feasibility and cost were now considered important, including a concern over environmental and social issues. These issues were brought into sharp relief in the 1973–74 war in the Middle East, the doubling of petrol prices, rising in inflation, worldwide recession, and increases in unemployment. Economic growth had declined and other quality of life issues were becoming important. The Economist Intelligence Unit had published one of the first critiques of transport studies in 1964. The EIU (1964) argued that forecasts had been based on the unconstrained use of the car in cities, but if new roads were not built to continue this level of car use then conditions would change and some of the forecast growth in traffic would not take place. The key point was that the level of traffic demand is dependent upon the levels of congestion and by implication on the costs of travel.

There was a period of constructive re-examination prior to a vitriolic attack on the use of the systems approach to the analysis of transport problems. It is interesting to note that a similar attack on the use of systems approaches for planning analysis had taken place both in the United States (Lee, 1973) and in Britain (Batty, 1976). The re-examination criticized the narrow conceptual base of the methods, the inflexibility in the basic structure of the process, the lack of any real theoretical development, and the fact that the models did not seem to be responsive to the changing demands of the politicians. Lee's (1973 and 1994) requiem for large-scale models suggested that the goals had never been achieved and often they had not been stated at the beginning of the study. For example, public transport had been promoted as a reaction against the impacts of highways and not as a real alternative in their own right. Lee also argued that for each objective offered as a reason for building a model, he could suggest a better way of achieving that objective (for example more information at less cost) or propose a better objective. It was a plea for the use of simple methods, for starting with a problem that needed solving not a methodology that needed applying, and for a balance between theory, objectivity and intuition.

In many cases it was accepted that the criticisms and proposals were valid, but it was also argued that the scale of the attack was unjustified. Kain (1978) stated that most criticism had taken place against some normative ideal. Criticism should be placed within the context of alternative problem solving frameworks. Most of Lee's comments were directed at the situation in the United States where the scale of the transport studies was considerably larger than in Britain, where order, model consistency and comprehensiveness in approach were more apparent.

However, apart from the changes in the economic climate and the move away from road building towards management solutions, there were concerns over the assumptions made in the studies, the scale of the approach and the output. The expectations of growth in population, employment, incomes and car ownership were all very optimistic, and the studies assumed that most growth would take place in the city centre, not at peripheral and greenfield sites. If the MALTS study is taken as an example it can be seen that population within the core was declining even before the study was completed. Between 1966 and 1971 the centre lost 34% of its jobs, a much faster rate of loss than in the rest of the conurbation. Car ownership growth assumptions were also in error as ownership levels reached one car for every five people in 1981, only a half of the forecast figure for 1991 and work trips declined by 10% from 1966 to 1971. If the basic assumptions were proving wrong even while the plan was being prepared, it is hardly surprising that the forecasts also turned out to be wrong.

Excluding the tunnel, the number of trips into the city centre fell by 7.6% between 1978 and 1983 whereas they had been predicted to rise in line with the projected increase in employment. The city centre had not become congested. Traffic speeds in peak hours actually rose during the 1970s, and off street parking in central Liverpool was only 65% occupied in 1981. Similarly, the expected growth in rail travel did not materialize. As elsewhere, the numbers of bus passengers declined, and in 1971 they were only just over half their 1961 levels.

More generally, the Merseyside pattern is found in other towns and cities. Evans and Mackinder (1980) carried out a survey of 31 early British transport studies to compare forecast change with actual change over a 10 year period. The error on average was 12% for population and employment, and over 20% for income and car ownership. They came to the conclusion that if no change had been assumed the results would have been no less accurate! There was some evidence that the models themselves had worked reasonably well but it was the prediction errors in the exogenous variables which had caused over prediction of the increases in demand. Probably more important than these technical reasons was the expectation among the politicians and local authorities that highway schemes should be built and that they were an essential part of the modernization process.

There was also a debate between those who were arguing for comprehensive long-term planning with clearly defined levels of investment and those who were promoting a more piecemeal approach to investment. This is best illustrated in the influential papers by Beesley and Kain (1964 and 1965) which criticized the comprehensive approach to transport planning proposed in *Traffic in Towns* (Ministry of Transport, 1963). They argued that the comprehensive approach

> misrepresents the problems attendant upon rising car ownership now, and in the coming decades, and much overstates the extent of the problem and the scale and kinds of policies needed to deal with it. This is unfortunate because, if anything, it reduces the probability that more investment will be prised from a reluctant Treasury, which is skilled at spotting flaws in advocacy. In concentrating attention on action necessary to meet a situation held likely to arise in 2010, the Report (Traffic in Towns) encourages easy acceptance allied with practical determination to do very little. More important, however, the Report fails to provide an alternative for action, even if; with the best will in the world, the level of investment expenditures available for urban improvement are much less than the Report's authors feel is desirable. (Beesley and Kain, 1964, p. 200)

Beesley and Kain suggest selective investment in existing primary routes near city centres together with some investment in new infrastructure. The expected growth in car ownership and use predicted in *Traffic in Towns* was exaggerated. Growth could be accommodated through limited investment and extensive traffic management and demand management, including parking controls and road pricing. With the benefit of hindsight, it is still difficult to say which view was 'correct', particularly as the counterfactual situation is unknown.

A further dimension of the rejection of the 1960s approach was the rise of the environmental movement in the early 1970s and the encouragement given by the Skeffington Report to the public and pressure groups to get more fully involved in planning issues. The general public had become much more aware of the question as to whether a road should be built, the weaknesses in established analysis and evaluation methodologies, and the presumption that redevelopment and blight were the inevitable consequence of progress.

The opening of Westway in July 1970 epitomized public concerns. This section of elevated motorway in West London passed through communities, very close to housing and seemed typical of the cavalier approach of highway engineers to existing structures (Van Rest, 1987). It was partly as a response to public concern about schemes such as Westway that the Urban Motorways Project Team was set up in 1969 to:

♦   examine current policies for fitting major roads into the urban fabric;

- consider changes that would improve the integration of roads with their environment;
- examine the consequences of such changes for public policy, statutory powers and administrative procedures.

However, even in 1969, the government reaffirmed its commitment to 'devote a substantial and growing part of the programme to road works in urban areas' (Ministry of Transport, 1969), and the 1970 White Paper on Roads for the Future required an investment programme of £4000 million (at 1970 prices) to be spent over 15–20 years. This programme was designed to 'check and then eliminate the congestion on our trunk road system' (Ministry of Transport, 1970). The main objectives were to:

- achieve environmental improvements by diverting long distance traffic, and particularly heavy goods vehicles, from a large number of towns and villages, so as to relieve them of the noise, dirt and danger which they suffer at present;
- complete, by the early 1980s, a comprehensive network of strategic trunk routes to promote economic growth;
- link the more remote and less prosperous regions with this new national network;
- ensure that every major city and town with a population of more than 250,000 will be directly connected to the strategic network and that all with a population of more than 80,000 will be within 10 miles (16 km) of it;
- design the network so that it serves all major ports and airports;
- relieve as many historic towns as possible of through trunk road traffic (Cullingworth, 1988).

The early 1970s marked a sudden change in priority brought about by events at all levels of decision-making. At the national level the Urban Motorways Comittee (DOE, 1972*b*) recommended integration in the design of roads with the surrounding area and inclusion of the costs of remedial measures as part of the total construction costs of a scheme. The report was published at the same time as a review of the land compensation aspects of road construction, and together they formed the input to the 1973 Land Compensation Act (Starkie, 1982). The second influential report was that of the House of Commons Expenditure Committee on Urban Transport Planning (1973) which reinforced the general trend away from investment in roads to better management and use of existing resources (Table 2.2).

With respect to urban transport planning this report marked a watershed as it summarized many of the concerns expressed over the 1960s approach to the production of traffic plans. These traffic plans had been used to test traffic

Table 2.2 Urban transport planning in 1973

Two main recommendations:
(a) National policy should be directed towards promoting public transport and discouraging the use of cars for the journey to work in city areas.
(b) Department of the Environment should have a positive policy for urban transport, laying down a broad approach which it should ensure that local authorities follow.

Detailed recommendations under four main headings – many have been implemented (about half by 1980 according to the Specialist Adviser).
(a) **Promotion of public transport** – including investigation of rapid transit systems, introduction of bus lanes and busways, implementation of bus management schemes, relief to taxis for VAT, operating subsidies for public transport and rates rebates for firms which stagger hours.
(b) **Traffic restraint** – including extension of parking meters, higher prices for parking, reductions in long-stay car parks, local authority control over off-street parking, maximum limits on car parking spaces with new commercial developments, keeper liability for parking offences, increases in the levels of fines for serious parking offenders, more widespread use of pedestrian precincts, routes for heavy lorries, the possibility of shifting taxation from the annual licence to petrol tax and a review of all road schemes in urban areas that had not been actually contracted out.
(c) **Transportation studies** – should still be carried out but with analysis of how people can be deterred from travelling and how many people might make use of improved public transport services. Alternative proposals should be significantly different from each other and the proportion of existing schemes considered as firm commitments should be minimized. The assumptions made and methods used should be more comprehensible to the layman.
(d) **Organization and financing of transport** – clearer allocation of transport planning responsibilities required for DOE. The urban roads programme should be replaced by an urban transport programme with greater emphasis being given to social research and analysis of public attitudes. New grants should be available for buses and expenditure programmes for urban transport should be based on definable and explicitly stated planning objectives.

Source: House of Commons Expenditure Committee (1973)

and transport alternatives within a single set of land-use patterns. Often the transport proposals tested were very similar to each other, and distributional issues were not investigated. The Select Committee recommended that transport studies should be carried out within a clearly defined framework of general policy. The bias towards roads investment and motorway construction was evident, and local planning authorities seemed prepared to accept the recommendations of the transport studies without a critical examination of the assumptions.

The main weaknesses identified were the inability to predict how many people would be deterred from travelling under certain restrictions and the inadequacy in predicting how many people might use new or improved public transport facilities. In most transport studies a large proportion of the budget had already been committed to urban motorway projects and many of these schemes had not been fully evaluated. In future a wider range of alternatives should be evaluated and the proportion of schemes considered as firm commitments should be kept to a minimum. Finally, it was proposed that alternative strategies should be considered at an early stage, and that the assumptions made and the methods used in transport studies should be comprehensible.

The recommendations made by the Select Committee summarized the growing dissatisfaction with the approach adopted and the priorities allocated to road construction in the 1960s. By capturing the public, the academic and the political concerns, it managed to achieve a switch in priorities away from road plans to the better management and use of the existing infrastructure through restraint on the use of the private car and through priority given to public transport. In terms of analysis methods it stressed the importance of evaluation, the strong links between land use and transport, and the issues which were of concern to policy-makers, in particular the modal choice decision.

At the local level political opposition to urban road building was growing. The most celebrated case outside London was the decision in Nottingham to abandon much of their highway construction plans and to look at the possibility of controlling traffic entering the city centre. The zone and collar scheme was introduced in the western part of the city on an experimental basis in 1975. Private cars would have limited access to the city centre and would be delayed by traffic lights at peak periods. Long-stay city centre car parking was reduced and prices raised. The intention was to get car users to switch to public transport for city centre trips and to use park-and-ride schemes. The Nottingham scheme was not a success and was abandoned shortly afterwards (1976). However, its significance was that it reflected public concern over urban road building and it demonstrated that an alternative did exist. But, it also illustrated the difficulty of introducing radical transport policies, particularly those which required people to switch from car to public transport. Getting people out of their cars has proved to be one of the hardest and most politically sensitive transport policy objectives. In retrospect, it can be seen that it was unwise to use time delay as a means to control traffic. Time is a resource which cannot be saved, so inefficiency was introduced into the system unnecessarily. It may have been preferable to have used a pricing mechanism to limit access, but this would have created its own problems, namely political acceptability.

In London, the London Amenity and Transport Association (LATA) produced its own report on Motorways in London (Thomson, 1969) which made a strong case against any urban motorways in the capital. This report coincided with the Greater London Development Plan inquiry set up to consider objections to that Plan. These objections exceeded 2,800 and mainly related to the transport proposals for a series of orbital motorways. The Panel recommended scaling down the orbitals to the M25 (substantially outside the GLC area) and the most controversial inner ring road. However, it was the political changes in London that marked the end of these urban road schemes.

The London Labour Party, after much debate, decided to fight the 1973 elections on an anti-motorway platform. This decision won them power and London's motorway plans were effectively abandoned for the next 10 years.

Although some cities, such as Glasgow, Leeds and (as noted earlier) Liverpool, maintained urban roads programmes, most cities abandoned their road building programmes, at least temporarily. The philosophy of the Buchanan Report which had been the standard for policy in the previous decade was also abandoned. Cities would no longer be subjected to substantial road building programmes to allow unrestricted access by private cars. In the Autumn of 1973, John Peyton, the Minister of Transport, announced in a short speech to the House of Commons 'that he proposed a switch in resources in the transport sector away from urban road construction.' This statement was made exactly 10 years after the publication of *Traffic in Towns* (Starkie, 1982).

The early 1970s marked the end of the aggregate large-scale land-use transport model. The doubling of world oil prices in 1973 and the impact that this had on world recession, inflation and unemployment, may have been the trigger to end large-scale urban road investment programmes. However, as noted here, this disenchantment was much wider spread, ranging from the political concerns over the costs and electoral implications of unpopular programmes, through technical concerns over the accuracy and validity of the systems approach, to the social concerns over who was actually benefiting from road building. All these events taken together meant that 1973 marked the most important watershed in thinking on urban transport planning.

The re-emergence of the land-use transport study: the 1970s

It may seem surprising after such a wholesale condemnation of the systems analysis approach to transport planning and its political and public rejection that there was a re-emergence of the approach as part of the new strategic planning system set up in the 1970s. Investment would no longer be argued on the basis of where to allocate economic growth, but that investment was now essential to prevent inner-city decline. Modern cities in the late twentieth century required high quality road and public transport infrastructure for their survival.

The 1968 Transport Act set up four Passenger Transport Authorities in the main conurbations (Merseyside, the South East Lancashire and North East Cheshire [SELNEC] area, the West Midlands, and Tyneside) to integrate all forms of transport and to coordinate transport with urban planning. West Yorkshire and South Yorkshire became two new PTAs after the 1972 Local Government Act. With the restructuring of local government all county authorities were required to draw up Transport Policies and Programmes

(TPPs) (operational from 1974) in which comprehensive policies for the development and operation of transport were presented. These annual plans were to cover roads, traffic management, parking and public transport, and their content should include a statement on local transport objectives, estimates of income and expenditure for the following financial year and a programme to implement the stated longer-term (10–15 years) strategy. The Department of the Environment (and the Department of Transport from 1976) would judge whether the preparation of the TPPs was adequate and the overall costs were within available resources. It would also ensure national and regional considerations were taken into account. The TPP was the basis on which support for transport from central government was distributed to the counties through the Rate Support Grant and the Transport Supplementary Grant (TSG was only about 4% of the size of the RSG). Local government reorganization, together with the new TPP system, was designed to replace the plethora of different authorities and agencies with less than 50 counties which would now be responsible for transport planning. As Starkie (1973) summarizes, these authorities would be adequate 'both in size and in resources to operate with a sensible external independence and internal cohesion, thus providing a stable environment for policies to take effect.' The specific transport requirements formed part of local government reorganization which also set up a system of structure and local plans. The TPPs should be linked in with the structure plan strategy and take account of the detailed implementation of policy outlined in the local plans.

Returning to Merseyside as a case study, the Structure Plan team did not attempt a systematic analysis for the options as had been carried out in the 1960s MALTS study. Instead, they took a wider perspective on the future of the conurbation and one based on much less data collection and analysis, but with little emphasis on the integration between transport and land-use planning. Although, on the land-use side, the Structure Plan's concern over inner-city regeneration might improve the prospects for public transport, nothing was said about how an increase in revenue support for public transport might be expected to make a positive contribution to that regeneration.

The influence of MALTS was still in evidence. Even though over 50 of the original roads schemes were abandoned, 17 were retained in the preparation pool and 10 were approved for construction before 1986. But more striking was the way in which they were justified. While some of the same arguments used 15 years earlier were still put forward, essentially the schemes were now seen as a means to promote economic development and traffic growth, not as a means to accommodate growth that was assumed to be inevitable.

Despite all the changes that had taken place in the planning assumptions, the inner ring road in Liverpool was reinstated. It received a tremendous battering at the examination in public of the Structure Plan, and was finally axed by the Labour administration which took over Liverpool in 1981. It seems that the same approach adopted in the MALTS study had re-emerged in the 1970s in a slightly modified form.

An examination of all conurbations would reveal many features in common with Merseyside. In the 1960s and early 1970s, large-scale aggregate models were used. In the late 1970s and early 1980s these gave way to a broader based approach to planning. At the same time, a policy based on the allocation of growth through dispersal had been replaced by one aimed at urban regeneration. Most remarkable is the apparent continuity of some of the key elements, highlighted in the Merseyside evidence (Banister and Botham, 1985).

- It took nearly 20 years to realize that city centre congestion was not a problem and that the present infrastructure was more than adequate.
- Transport policies were still thought to be effective instruments of land-use policy. In the 1960s that faith was attached to road building, but in the 1970s it was attached to low fares.
- Methods of analysis had not improved. The MALTS methodology made what can be seen to be serious mistakes, because the assumptions on which it was built turned out to be wrong. The Structure Plan process which followed it aimed to avoid what was seen as the inflexibility of the old style plan. But in the end it undertook no analysis to justify those transport policies which were being promoted.

During the 1970s the re-emergence of transport planning was a continuation of the 1960s approach, but under the guise of the Structure Plan and to a lesser extent the Transport Policy and Programmes. The analytical skills were still required, but more in the planning context, to predict changes in population, employment and local economic activity (Batty, 1976). However, it was during the 1970s that several other transport policy issues required the attention of researchers. The common factor underlying each issue was the new policy imperative to ensure value for money in times of reduced public investment, and to make the best possible use of existing infrastructure and resources. These concerns manifest themselves in a variety of ways, five of which will be highlighted here.

Traffic management

Traffic management required the best use of the existing infrastructure through traffic control, bus priority schemes and the regulation of the supply and price of public car parking space. The government's Transport and Road Research Laboratory carried out a series of experiments on urban traffic control including the Leicester (1974) purpose built automated traffic control system. The adoption of bus lanes was less than enthusiastic in the early 1970s and parking controls seemed to lack effectiveness, as the costs of collection were 50% higher than the revenue recovered in 1972–73 (costs of £48 million and revenues of £32 million). The most radical proposal was the feasibility study carried out in 1973 on area licensing in central London, where a supplementary charge would be made on cars and other vehicles entering the licensing area. In 1974 the technical officers reported favourably but the new Labour GLC decided not to take the scheme any further.

The roads programme

The roads programme in the 1970s was severely curtailed as the government increasingly emphasized coordination in transport planning through the TPP circulars, and this imperative was reinforced by a review of all major road schemes in each county's programme which were not due to start before April 1978 (Circular 125/75, Department of the Environment, 1975). The government's expenditure cuts in the 1970s fell heavily on the roads programme. When the White Paper on Transport Policy was published in 1977 (Department of Transport, 1977a), it was revealed that road construction in progress on the English motorway and trunk roads programme had contracted from about £400 million in 1971–72 to about £270 million in 1976–77 (1970 prices). Thus new trunk road building had contracted by 32% but local authority road building by 50% (Van Rest, 1987).

Many of the ambitious highway plans recommended by the large-scale aggregate transport studies of the 1960s and early 1970s were axed. Most of them were behind schedule, and economic recession and the restructuring of industry had led to a decline in inner-city employment and pressure on peripheral development, often on sites made accessible by new motorways. The roads that were actually built seemed to fit in more closely with other priorities such as comprehensive traffic management schemes.

The environment

The environment was also an issue which had raised public concerns, both in terms of the increased land take required for new roads and urban development, and the externalities created by noise and pollution from road traffic. The benefits of a high speed motorway and primary road system for the user were being balanced against the loss of land and a general reduction in the quality of the environment for others. A series of reviews and measures were taken, principally against the lorry, through the Armitage Inquiry (Department of Transport, 1980) and the Foster Inquiry (Department of Transport, 1979b). The size of lorries proved to be one issue upon which the environmental lobbies influenced policy. For a decade they were successful in delaying and blocking the powerful industrial interests who wanted an increase in the gross weight and dimension of lorries (Starkie, 1982). It was only at the end of the decade that the argument for heavier and larger lorries was won, and even then it seemed that this victory was achieved because of recession and low economic growth brought about by the second oil crisis in 1979, rather than the actual merits of the case. In times of economic recession and hardship, environmental issues become of secondary importance. A lesson which may still be true two decades later!

The government proposed that the resolution to the lorry problem would be the construction of a lorry network, but this proposal in turn became a non-event as the whole roads programme came to a halt when public expenditure was cut back. However, local authorities did have powers to restrict heavy lorry traffic (over 3 tonnes unladen weight) to particular routes 'so as to preserve or improve the amenities of their areas' (Heavy Commercial Vehicle (Controls and Regulations) Act 1973). In both the cases of lorry weights and sizes, and of lorry routes, the arguments between reductions in hauliers' costs and environmental benefits lay at the root of the disagreements between the industry and environmental lobbyists. More fundamentally, the 1970s marked a transition in transport planning analysis. No longer was the analytical expertise of the technocrat seen as unquestionable. There were as many experts arguing against a scheme as for a scheme, and for the first time evidence, which had previously been accepted, was now being placed under the microscope.

The public inquiry

The public inquiry was the forum within which many of these debates were held. As with the rejection of the 1960s approach, concern with the process of consultation and participation was at first muted, but soon it became a full scale confrontation between the opposing interests. There were several reasons for this discontent. The government maintained that the public inquiry could not question public policy on roads or the basis on which traffic forecasts were made to justify the schemes. The protestors claimed that they should be able to question the need for a road, and they were also concerned about the independence of the inspector (appointed by the Secretary of State for the Environment and Transport) and the treatment of those protestors with legitimate cases at the inquiry, including the availability of information to them and the costs of being represented at an inquiry. Their success was not so much in getting decisions reversed, but in raising public concerns over roads, in gaining publicity for their cause, and in delaying the road building programme. It was now taking over 12 years for a road to complete the process of preparation, inquiry and construction.

Two major initiatives were set up by the government in the mid-1970s. In December 1976, the Advisory Committee on Trunk Road Assessment was set up to review the Department of Transport's Trunk Road Assessment methods, including the procedures used for traffic forecasts and the relative importance attached to economic and environmental factors. The second initiative involved the procedures surrounding the conduct of public inquiries into highway proposals (Department of Transport, 1978a, 1978b). Certain changes were recommended including the appointment of inspectors by the Lord Chancellor rather than the government department, and the introduction of a pre-inquiry procedural stage which would allow all information to be pooled and permit the basic programme of the inquiry to be agreed. Objectors were still not allowed to question government policy, but they could debate whether national forecasts were appropriate to that particular local situation or whether there were factors which might reduce or enhance the national forecasts.

It should be noted that statutory objectors who were likely to be materially affected by the proposed road scheme had a right to be represented at the inquiry, but that other objectors could only attend at the inspector's discretion. This distinction between the two basic types of objectors had been introduced in July 1976, just after the review of the public inquiries into highway proposals was set up. These two actions effectively diffused much of the opposition to the roads proposals, but, as noted earlier, the roads programme itself had been severely reduced through the public expenditure cutbacks of the 1970s. In future the government would also publish its roads programme as part of the new openness on information (Department of Transport, 1978a). Gradually, the confrontation between the various interest groups in challenging government policy and forecasts at public inquiries was dissipated. Priorities for road investment were the completion of the primary network of long distance motorways and a switch of funding to small town bypasses. The Policy for Roads document was published regularly and a layman's guide to the role of public inquiries into road proposals was also produced (1981). The temperature of the debate was again raised in February 1980 when the Law Lords overruled Lord Denning's ruling that objectors could cross-examine official witnesses on how they had calculated traffic forecasts. The Law Lords ruled that local inquiries were not the appropriate place to challenge the forecasts.

The government had been extremely successful in dissipating many of the legitimate concerns of the public through the introduction of a more sensitive and open procedure for the public inquiry, but probably the main reason was the end of the 1970s investment programme in new motorways. The rationale for protest had disappeared.

Public transport

Public transport patronage during the second part of the 1970s continued to decline with the bus now accounting for 12% of travel (Department of Transport, 1976*a*) and an increasing level of public subsidy. Total support to the bus industry more than doubled in just 2 years to 1975–76, with the bulk of subsidy going to metropolitan areas. It seemed that persuading people to use public transport in preference to their cars was really an impossibility. Public transport was increasingly only being used by those people without access to a car and the laudable aim of getting people to switch modes was unrealistic. The questions posed by the House of Commons Expenditure Committee (1973) and Table 2.2 have been answered, namely that there are no methods available to predict modal shift back to the bus unless a comprehensive approach to planning is adopted to give buses exclusive rights of way in cities. It is not just a question of generalized cost differences between the alternative modes, but the quality of the service offered by the bus. It has taken another decade of decline to realize fully the implications of this problem.

Even after the 1970s bus travel continued to decline, with reduced levels of service, increases in costs of operation and increases in fares which rose by 30% in real terms (1972–1982). Revenue support had also grown from £10 million in 1972 to £520 million in 1982, a thirteen fold increase in real terms (Department of Transport, 1984). Many conurbations had increased their support for public transport, mainly through maintaining a low fares policy. In Merseyside, ridership has increased from a low point of 216 million bus trips in 1981–82 to 254 million in 1984–85, with revenue support for buses rising from £11.8 million in 1981–82 to £18.2 million in 1984–85 (1975 prices). In an attempt to defend their position, the County's 1985–88 TPP submission quoted the Structure Plan panel to emphasize its case

> we believe that an effective public transport system will contribute to regeneration because it will broaden the range of accessible job opportunities, especially for the less well off, and will extend the catchment area within which an enterprise on Merseyside can recruit . . . Our opinion is that failure to maintain an effective public transport system will hinder the fulfilment of the strategy and will do so more grievously than a lack of resources to carry through new road construction.

The significance of being able to quote the Structure Plan and report of the panel presiding over the examination in public is that the government had accepted the plan in 1980. The County's submission went on to note that

> the value of this turning of the tide (in patronage) for the urban regeneration strategy needs to be measured in terms of the social, economic and environmental needs addressed in the Structure Plan's policies.

But the County Council had made little attempt to quantify the projected impact of cheap fares, nor did it monitor their effect on urban regeneration, once they were introduced. Thus, despite all the claims put forward for the policy, including its redistributive effects, there has been little attempt or empirical evidence to justify any of them.

The Merseyside policy outlined here was stated in the mid-1980s. This in itself is strange as the Conservative government had already made it clear in a series of Transport Acts (1980, 1981, 1982, 1983) that public expenditure would be reduced and that subsidy levels could not continue to increase. The thinking in Merseyside was more appropriate to 1975 when the Labour government was preparing its major piece of legislation on a social transport policy (the 1978 Transport Act).

This major statement of transport policy (Department of Transport, 1976a, 1977a) identified three policy objectives:

◆   to contribute to economic growth and higher national prosperity, particularly through providing an efficient service to industry, commerce and agriculture;
◆   to meet social needs by securing a reasonable level of personal mobility, in particular by maintaining public transport for the many people who do not have the effective choice of travelling by car;
◆   to minimize the harmful effects, in loss of life and damage to the environment, that are the direct physical result of the transport we use.

Other secondary objectives emphasized the efficient use of resources, notably energy, the importance of choice and local democracy, the interests of transport workers, and the need to restrain public expenditure. The well-established objective of providing an efficient transport system, that would contribute to economic growth and higher national prosperity, was supplemented by the social objective of meeting transport needs.

The county councils were also given stronger powers to coordinate transport and a requirement to prepare a County Public Transport Plan (PTP) to cover public transport within the context of the TPP. The plan should set out the policy for public transport in the county for 5 years ahead, and assess the transport needs of those people in the county and how those needs could be met.

Towards the end of the 1970s a considerable amount of research took place to determine how needs could be measured. This included an assessment of the different approaches adopted by the shire counties (Banister, 1981). Three basic methods were used – minimum levels of service or standards, demand measures, and local-based measures – but none was satisfactory either theoretically or empirically, and an operational tool for transport planning was still sought. The concentration on the identification and measurement of need may be misplaced, and needs can only be considered at the conceptual level with the aim of providing a perspective within which policy-makers can consider the distributional effects of particular decisions (Banister, 1983). The question of transport needs was never satisfactorily resolved, and with the switch from social concerns to those of identifying markets for transport, economic concepts of demand and willingness to pay again came to the fore in the 1980s.

Conclusions

During the 1960s and 1970s, nearly 2300 kilometres of motorways and 2500 kilometres of dual carriageway all-purpose trunk roads had either been constructed or upgraded, and the strategic network was near completion.

In future, priority would be given to specific projects, such as the M25 route round London, and it was recognized at last that transport planning should be more closely related to land-use planning. The assumption of a continued supply of relatively cheap fuel was no longer realistic, and decisions should now be made that integrated land-use functions, so that the absolute dependence on transport and the length of numbers of journeys could be reduced.

Over these two decades there seems to have been a continuity of roads policy (Table 2.3) which was only terminated with the 1978 Transport Act when social objectives and commitment to public transport were added to policy concerns over the efficiency of the transport system. Even then this 'end' of road building was only temporary as the radical Conservative governments of the 1980s changed both the focus of transport policy and the

Table 2.3 Continuity of roads policy in the 1960s and 1970s

1946	Labour	Adopted the 'tea-room' plan for trunk roads
1951	Conservative	Continued with that plan
		Developed urban road traffic policies
1964	Labour	Continued with the 'Marples' policies
		Revitalized the trunk roads programme
1970	Conservative	Implemented this strategy
		Developed urban transport policies
1974	Labour	Adopted the urban policies
		Reappraisal of trunk roads policy
1979	Conservative	New generation of motorway construction

Note: The tea-room plan was produced in 1946 by Alfred Barnes, the Minister of Transport. It is said that these plans were exhibited in the Member's tea-room at the House of Commons and that the plan looked like an 'hour glass' shaped network of motorways covering some 800 miles , and it was the intention of the Labour government to complete this plan in 10 years (it was only in 1971 that the first 800 miles (1280 km) of motorway were opened in Great Britain)

Source: Based on Starkie (1982) and British Road Federation (1972)

frequency with which major policy changes took place. Trends in the 1960s and 1970s in the economic and social cycles were fairly smooth with more abrupt changes brought about by the two 'oil crises' of the 1970s. But transport policy demonstrated 'distinct, quite sudden but infrequent change' (Starkie, 1982).

These changes usually took place towards the end of a particular administration and the continuity was supplied by the next government. Starkie (1982) argues that the civil servants wished to avoid conflict and seek consensus, and this strategy deflected challenges to existing policy. If a sufficient consensus against existing policy fails to focus upon a generally agreed new course of action, policy regresses towards a non-policy (Plowden, 1973), such as the switch away from urban motorways to managing demand.

Chapter 3

Developments in planning analysis and evaluation

Parallels in land-use analysis

Over a similar period of time, an almost parallel development of analysis processes was taking place in land-use and urban systems analysis. Again there seemed to be a natural progression from the realization of the complexity of urban problems and the use of the scientific method for the analysis of social problems towards a systems view of the way in which cities and regions were being perceived. Throughout their development, land-use models have been linked directly to public policy, hence the precarious nature of their existence. There is a compromise between the theoretical acceptability and their practical feasibility (Batty, 1981).

The years 1960–1965 saw an explosion in land-use modelling. These early models were based on fairly strong assumptions and were concerned with the representation of the city. As such they were aggregate in nature and vast in scale. The expectations were that modellers and planners would produce tighter and stronger answers to urban problems, but in reality there was a mismatch between these expectations and the realization of the vast size of the task. At the end of the 1960s there was a fairly general feeling that the models had been oversold and the initial optimism was fading (Butler *et al*, 1969).

Three particular problems seemed to summarize the difficulties. Many of the models were never actually built because of their size and complexity. Secondly, the models did not explain phenomena of interest to decision-makers and this was also reflected in the ludicrous predictions of growth in urban areas. Finally, the organizational environment was not conducive to the development work necessary to make the models operational. The analyst often had to learn on the job about the problems and there were long periods with no apparent progress. The decision-makers wanted instant answers from instant models. More fundamental though was the absence of any coherent theory of how urban systems worked, and the methods themselves had often been taken from other disciplines, in particular the physical sciences, operations research and statistics.

The gentle critique soon developed into a full scale review. It was started by Boyce, Day and McDonald's review (1970) of metropolitan plan-making in 13 major urban areas in the United States. They examined the approaches adopted by planning teams in preparing and evaluating alternative strategies, and found that there had been too many abortive data collections on a massive scale, that too much time had been spent on the development of techniques and not enough time on the evaluation of alternatives, and that in most cases the planning process was conceptualized as a linear sequence leading to the selection of one plan from a range of alternatives. There were also more general attacks on the whole field of systems analysis and its use in public policy (Hoos, 1972; Berlinski, 1976) and on particular applications of land-use models. For example, Brewer (1973) reviewed the mismanagement of the San Francisco and Pittsburgh Community Renewal Programmes, both in terms of the methods used and the organization of the studies themselves. The culmination of the review was Lee's (1973) requiem for large-scale models.

In almost every case it seemed that the time, the data and the expertise were not available to develop a general land-use and transport model which was consistent in its level of detail. The only exception was Lowry's model developed for the Pittsburgh Regional Economic Study (Lowry, 1964). This model simulated urban development, given information about the spatial distribution of basic employment and the costs of travel. The model never achieved popularity in the United States, but was used in the early sub-regional studies in Britain (Batey and Breheny, 1978).

The 1970s marked the revival of systems analysis, with more theoretical work and more modest claims for what they could and could not achieve. Pack's review (1978) of work in the United States demonstrates that there was a proliferation of modelling applications in the 1970s. The use of more transparent methods, better communication and the breaking down of barriers between planners and decision-makers all facilitated this process. One of the most effective arguments put forward by Kain (1978) was that most of the critique had been made against some normative ideal. Effective criticism must be based on alternative ways of tackling urban problems. He cited the Community Renewal Projects in the United States, commenting that there were several hundred of these, but it was only those in San Francisco and Pittsburgh that had any influence on housing research and policy. These were the two which had used land-use modelling techniques.

Land-use activity allocation models capable of estimating future patterns of land development and transport flows were the main contribution of planners. Much of the thinking was based on the Lowry model (Lowry, 1964) which split employment into two categories, with basic employment being allocated exogenously and service employment being allocated by the model to reflect the distribution of households and employment, both of which were calculated within the model. Other similar approaches all used the gravity type allocation model to link the mechanisms of urban growth and change (Hutchinson, 1981).

Models which calculate urban activity allocations are very dependent upon the land development constraints established exogenously. The main determinant of the levels of interaction is the cost deterrence function, and once the existing activity has been reproduced, the assumption that this function remains stable is crucial to future activity allocation. The impact of changes in the land market through land availability and rent changes, and the advantages of different activities locating in close proximity to each other (agglomeration economies) have not been fully developed.

In Britain, the motivation was similar to that in the United States but the underlying structure of planning was different. Until the 1960s planning in Britain was dominated by architects and physical planners. It was only at the end of the 1960s that a fundamental change took place with the introduction of social sciences to planning and the adoption of the systems approach to analysis (McLoughlin, 1969). The first urban models were built in Leicester, Manchester, Bedford and Cambridge and they were based on shopping models applied in Haydock, Oxford and Lewisham (Foot, 1981). The links between land use and transport were not as strong as in the United States, and most were built on the sub-regional scale. This first generation of models was designed to allocate growth and the analysis was often carried out by small teams working outside the constraints of the local authority. Outputs were specific with the emphasis on order rather than the US tendency for comprehensiveness and size. Systems analysis was also concerned with comprehensiveness and the rational planning process. Planning methods such as Lichfield's Planning Balance Sheet Appraisal and Hill's Goals Achievement Matrix (GAM) were developed for evaluation and to supplement the analytical methods such as gravity models, economic base techniques, Lowry models, potential surface models, and cohort survival methods (Lichfield, 1970; Hills, 1973).

The growth in the popularity of the systems approach coincided with the reorganization of local government (1974) and the introduction of structure and local plans. In theory this new environment should have marked an increase in the use of models, but the reverse occurred. Structure plans were established in the 1968 Town and Country Planning Act to replace the existing system of development plans which were mainly zoning exercises. The new system introduced two tiers of plans (Table 3.1). At the strategic level, the structure plans were designed to state the broad policy objectives for the counties, with local plans concerned with the detailed implementation of policy, in particular development control. On the whole the structure plans, with a few notable exceptions (South Hampshire and Teeside), failed to achieve the level of technical sophistication evident in the sub-regional plans developed in the late 1960s and early 1970s (Batey and Breheny, 1978).

Until the 1970s the accepted means of plan-making had been to survey the

Table 3.1 Development plans to 1986

Major changes to the planning system introduced by the Town and Country Planning Act (1968)

Structure plans: Produced by metropolitan counties and shire counties for ministerial approval. Statement of a broad strategy for development including associated transport systems and broader social and economic objectives.

The last of the 82 first generation structure plans in England and Wales were approved in July 1985. By April 1986 there were three replacement structure plans and 30 alterations to original plans which had been submitted but not yet approved.

It took 14 years to complete the first cycle. Between 1981–1985 there were 32 approvals with an average time of 28 months from submission to approval. The delay was even longer if public participation and preparation time were included. These very lengthy documents were often out of date when approved.

Local plans: Produced by local authorities for adoption. Formulated detailed proposals for the development of land including maps. Set out development control policies either for a limited area of comprehensive change (action areas) or for larger localities (district plans). Subject plans dealt with specific policy issues such as housing, minerals or conservation.

By March 1986, 474 local plans had been adopted and a further 269 were on deposit prior to adoption. It was claimed that many were too detailed and that delays in approval of structure plans had affected local plans.

area, to analyse its known or anticipated problems, and to produce a plan to describe a state of affairs expected or desired at some future date – the classic survey, analysis, plan sequence. The new system of plan-making accepted a process that was continuous, cyclical and took a systems analysis view based on the identification of needs, the formulation of goals and the evaluation of alternatives, with subsequent monitoring and modification of adopted plans.

Again, as in the United States, there was a reaction against systems analysis in Britain. It has been suggested by Batty (1976) that this was part of the action-reaction cycle for and against models and their use in planning. In 1967 land-use models were first used in Britain and optimism was high when, in 1970, the Department of the Environment issued guidance on the use of predictive models in structure plans. By 1974 the reaction was already setting in with further Department of the Environment instructions (DOE Circular 98/74) suggesting that counties should concentrate on key issues in their structure plans. As with the transport models, there was an increased interest in the economic and social aspects of planning, and a growing scepticism about the ability of planners to produce long-term physical plans. Breheny and Roberts (1981) concluded that much forecasting work was weak, both in terms of the way forecasts were used and the way they were produced. In particular, their review of 13 of the 60 available structure plans criticized the weakness between forecasting and policy formulation.

The mid-1970s coincided with a downturn in anticipated growth levels as an outcome of the oil crisis (1973–1974) and a move away from the rational comprehensive approach to plan-making towards a pragmatic incremental approach (for example East Sussex). With the introduction of the structure planning system there had been a separation of powers between the new local government counties and the districts. Implementation was at the local level and so structure plans were seen to be largely ineffective, particularly as the process of submission, inquiry and adoption was often lengthy. With the switch from growth to decline, planners had a much harder task. Allocation of growth may be difficult but accommodating decline is much harder. Urban modelling had not suggested that inner-city decline would take place together with the massive decentralization of activities and changes in the employment structure. The priorities for the planning process had changed from forecasting growth towards an emphasis on the short term and the need for flexibility.

Towards the end of the 1970s, planning was changing with the move away from large-scale publicly funded urban renewal schemes towards giving additional powers to local authorities with severe inner-city problems so that they might participate more effectively in local economic development. The Inner Urban Areas Act (1978) set up partnerships (7) in the most severely deprived areas, together with programme authorities (23) and other designated areas (16) in locations with less severe problems. An inner-area programme was to be produced as a bid for funding, and it was to involve all parties – private and public, statutory and voluntary.

This move towards a greater involvement of the private sector was continued after the General Election of 1979 with an extensive urban regeneration programme. Planners were becoming negotiators between the private and public sectors, and their role was no longer primarily regulatory but more development based, seeking opportunities for planning gain (Savitch, 1988). The Urban Development Corporation was the main means by which property investment could be stimulated. The responsibilities of local authority planners were pre-empted as the role of the UDCs was to make sites available to developers and to provide the appropriate infrastructure so that investment would follow. It was ironic that one of the most interventionist actions concerning urban planning was taken by a government which was supposed to allow the market to operate (Lawless, 1986).

The 1980s have provided a major paradox in UK planning, with strategic planning becoming increasingly unfashionable in both central and local government, yet many strategic planning issues have captured the headlines (Breheny, 1989). The concern over long-term development and environmental quality had been replaced by short-term gains and negotiations. The new definition of planning was the process by which government enables the private sector to invest profitably in urban areas. The earlier construction of urban problems as defined by poverty and inner-city decline was reformulated in terms of competitiveness and fiscal solvency (Smith, 1988). The government's intentions were made clear in the White Paper 'Streamlining the Cities' (1983) where it was stated that the earlier confidence in strategic planning has been 'exaggerated' (para 1.3) and the search for such a strategic role may have little basis 'in real needs' (para 1.11). The abolition of the strategic level of government in London and the Metropolitan Counties took place in 1986 (Table 3.2). This was followed by the proposed abolition of structure planning in the Consultation Paper, 'The Future of Development Plans' (September, 1986). This proposal was formalized in the White Paper of the same title (January, 1989), but no action was taken. As Breheny (1989) observes, planning at the strategic level and many of the administrative structures which were designed to support it have been largely dismantled as part of the continuing process of streamlining the administrative procedures surrounding land-use planning.

Table 3.2 Development plans since 1986

Major changes to the planning system introduced in the Local Government Act 1985

1. **Strategic planning guidance** from the Secretary of State which was not formalized or centrally controlled but worked out in conjunction with local planning authorities with a draft for public comment.

2. **Strategic policies** to be established by the counties to guide the detailed preparation of development plans by the district authorities. The strategic policy to cover:

   • strategic highways and transport
   • policies on control of mineral operations and waste
   • provision of land for housing
   • policies on major retailing and industrial development
   • protection of the countryside

The county responsibilities also covered survey, analysis and the preparation of information, the process of consultation before publishing draft statements, the provision of an examination in public after consultation, and the publication of the adopted statement. There was no longer a requirement for the Secretary of State's approval.

3. **Unitary development plans** (for London boroughs and metropolitan district councils) and **District development plans**. These detailed documents are similar to local plans and cover all land-use policies (except minerals). They identify proposed locations for development.

   • Part I should contain general policies on strategy
   • Part II should contain more detailed proposals on land use and development
   • Both parts should be prepared together

Draft plans are published with a period of six weeks for representations and objections together with consultations with counties and the government. A public local inquiry would be held with a planning inspector to give an independent commentary on the plan. A plan is then published with modifications and adoption takes place after a further six-week period for comment. The adopted plan does not require the Secretary of State's approval, but he would receive a copy.

The new Unitary Development Plans (UDPs) were to embrace Planning Policy Guidance notes (PPGs), the Strategic Planning Guidance (SPG) prepared for each of the former metropolitan areas, and any special government designated urban policies such as Enterprise Zones and Urban Development Corporations. The new pragmatism and highly centralized planning process replaced the lengthy and data intensive strategic planning process by brief regional guidance notes from government, supplemented by strategic policy statements produced by the counties (Table 3.3). As early as 1986, the Secretary of State for the Environment announced that structure plans would be abolished as they were 'unwieldy and took too long to prepare'. It was only in 1989 that the White Paper proposed the replacement of structure plans by county statements which would give a broad outline of policy with details to be filled in by district development plans. These proposals have not been implemented but structure plans now no longer need Department of the Environment approval. They still have to go through public consultation and examination in public, and there is a requirement to cut out detail and to concentrate on key strategic decisions.

It is also ironic that as strategic planning disappears from public sector planning it is making an appearance in the private sector. Many large corporations are now committed to forward planning to assist with more rigorous decision-making (Breheny, 1991). This interest is particularly apparent in the retail sector where several of the major companies have set up specialist groups or employed consultants to carry out systematic analysis to identify suitable locations for new outlets and to assess the options for new forms of distribution. Many of the techniques developed in strategic planning (for example gravity models and operations research techniques) are being used. In addition, many private corporations would like to operate within a strategically planned environment as this reduces uncertainty and hence their

Table 3.3 The Patten model[1]

Regional scale:	County authorities form themselves into appropriate regional groupings to produce statements of regional advice (10–15 year period) which are then submitted to the Secretary of State for the Environment. Regional guidance for each region is then produced.
Structure plans:	Structure plans will continue to be legislatively based but will be self-approving through consultation processes. The new structure plans will not have to be submitted to the Secretary of State, but there will still be reserve powers to intervene. The new procedures will reduce time for plan preparation and allow more local initiative.
Local plans:	No change here except that the plans are now district wide. They are still legally required and must be self-approved.

Note
1 Chris Patten was Secretary of State for the Environment at this time; he then became Governor of Hong Kong and is currently one of the EU Commissioners
Source: Based on Breheny (1991)

own risks. Although many businesses may support the principles of free competition, when it comes to their own business most of them do not welcome competition, particularly if it is perceived to be unfair. For all its faults the planning system has provided both continuity and stability, and each entrepreneur was fully aware of the procedures to follow. It is only at the very localized level that planning controls have been retained. At the strategic level for major investment, such as a new settlement or a regional shopping centre, there is less confidence in this system of limited strategic guidance and the Patten model (Breheny, 1991).

By the end of the 1980s, all the Strategic Planning Guidance (SPGs) for the former metropolitan areas had been published. But this was immediately superceded by a new form of Regional Planning Guidance (RPG), introduced to supplement existing national planning policy guidance (PPGs). The RPG documents were produced for each of the eight English regions over the period from 1993 to 1996. In the meantime, planning was based on UDPs and the SPGs from the late 1980s, and it was really only at the UDP review stage that the new regional guidance was taken on board. The first UDPs were adopted in 1993, and 58 (out of 69) had been adopted by January 1999 (Thomas and Roberts, 2000). The non-metropolitan counties retained the two tier development plan structure of slimmed down structure plans for strategy and local plans for implementation. The move towards regional institutional building was initiated through the creation of the government offices for the regions (1994) to coordinate economic development and regeneration, transport infrastructure and land-use planning.

The Regional Development Agencies again strengthened regional government (DETR, 1997b) through partnerships with business, through regional conferences, and through a unification of the process by which regional planning advice would be produced. More specifically, they were charged with:

- environmentally sustainable development policies – the Regional Sustainable Development Framework – a non-statutory vision of how the region will contribute to sustainable development at the national level;
- integrated regional transport policies – the Regional Transport Study;
- achievement of regional economic growth strategies over a 10 year period – the Regional Economic Strategy;
- developing a renewable energy strategy – the Regional Renewable Energy Capacity Assessment.

The regional planning body is now responsible for preparing and updating the RPGs, and in the longer term may be an elected body with mayoral powers such as those introduced in London. The complexity of regional planning in the UK is becoming legend, and this power has been allocated without any real form of participatory democracy, and a similar pattern has been observed in several European countries (Motte, 1997). The new powers for the regions make an important contribution to the objectives-led planning approach that has evolved in England over the last 10 years (Figure 3.1). However, there is a danger that the economic priorities of income generation, reduced levels of unemployment and inward investment into the regions will take an overriding precedence over other environmental and social objectives. The vision for each of the regions must embrace all three elements within the framework of sustainable development (DETR, 1999a).

An integral part of the RPG is the Regional Transport Strategy (RTS) which sets out the regional transport policies and objectives, including the priorities for investment in and management of transport and infrastructure across all modes. It has a wide ranging brief, and includes parking standards, demand management, transport accessibility criteria and measures to improve transport choice (see later section on PPG13). The intention is to make policies more 'consistent' across regions and to view transport issues in a truly multi-modal perspective, rather than as a set of competing modes. It begins to address the question of how we can make the best use of the transport system that is already in place.

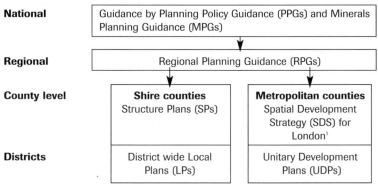

Note
1 In London, the Mayor is responsible for producing a Spatial Development Strategy, but elsewhere
   it is absorbed into Regional Planning Guidance

Figure 3.1 The Strategic Planning Framework in England since 1995 (Source: Based on Thomas and Roberts, 2000)

Evaluation in planning and transport

Cost benefit analysis

Across all transport and planning analysis the issue of evaluation has always proved particularly controversial. Ever since Dupuit (1844) concluded that more general benefits accrue to society than is apparent in revenues, decision-makers have been searching for techniques that can include the full range of factors in one analysis. It should be remembered that the basic purpose of evaluation is to assist the decision-makers in choosing between alternative courses of action. Early assessments were made on operational criteria such as whether it was physically possible for roads to accommodate traffic without unreasonable delay. Questions such as design standards, gradients and alignments were all important as were the costs, scale and timing of the improvement.

[Cost benefit analysis (CBA) can be defined as 'a practical way of assessing the desirability of projects, where it is important to take a long view (in the sense of looking at repercussions in the further as well as the nearer future) and a wide view (in the sense of allowing for side effects of many kinds on many persons, industries, regions, etc)' (Prest and Turvey, 1965).] Social cost benefit analysis (SCBA) goes beyond the consideration of private costs and benefits to the individual enterprise and covers a range of social costs and benefits. Welfare economics argues that within a general equilibrium situation, consumers exercise their preferences to consume precisely that combination of goods and services which best suits them. Producers respond to these actions by providing those goods and services which best meet these demands. Through the operation of the capital and labour markets this process is simultaneous with the selection by firms of the optimum technical processes, and the optimum combination of labour and capital inputs. Workers can balance their time between work and leisure activities. The distribution of earned income is determined by the interplay of demand and supply, and the distribution of wealth and unearned income is the outcome of savings behaviour and ownership patterns. Welfare economics attempts to accommodate some of the weaknesses of general equilibrium theory as well as extending it to cover some of the income distribution questions. However, many problems still remain such as the treatment of value, market imperfections and the treatment of consumer preferences. Some commentators (for example Dunleavy and Duncan, 1989) dismiss cost benefit analysis as a neo-Keynesian device for boosting public expenditure.

With respect to transport evaluation, cost benefit analysis has been very influential from its first application to the London–Birmingham motorway to the present time (Coburn, Beesley and Reynolds, 1960). Losses in travel times and increases in accidents were calculated from the case where the decision was not to improve the existing route to motorway standards. This basic approach was superseded in 1972 by the COBA cost benefit analysis method of appraisal (DOE, 1972a), where the costs and benefits of a new road were discounted over a period of 30 years from the proposed opening date to give a net present value and a net present value cost ratio. [It was solely an economic appraisal method involving savings in travel time, savings in vehicle operating costs and lower accident rates, which were matched against the capital and maintenance costs of the new road.)

[Social cost benefit analysis was first used in transport evaluation for the Victoria underground line study. In this application, the strict interpretation of user costs and benefits was extended to include the reduction in journey time and costs; the improved comfort and convenience experienced by those travellers diverting to the new line, the time and cost savings to traffic which remained on parallel routes; and the benefits to generated traffic on the Victoria line (Foster and Beesley, 1963).] This broader based evaluation transformed an expected loss of £2.14 million per annum to a healthy rate of return of 11.3% (with a discount rate of 6%), with over half the total benefits accruing to traffic which did not even use the new Victoria line.

By the middle of the 1960s cost benefit analysis was a common prerequisite of project approval, with a minimum rate of return of 8% required for public investment. However, it was at the end of the decade that SCBA had its greatest test with the 2½ year Roskill Commission Inquiry into the location of London's Third Airport (1968–1970). The scale of the inquiry and the thoroughness of the methods used made this the pinnacle of rational comprehensive planning. It was at this point that reaction against and rejection of this approach to planning and transport analysis took place. This change was most evident in the Greater London Development Plan which followed immediately after the completion of the Roskill Commission Inquiry (with the Inquiry lasting for 2 years from 1970 to 1972).

The Roskill Commission Inquiry into London's third airport

The Commission started by examining a list of 78 possible sites, and within 7 months had reduced this list to four: Thurleigh (near Bedford), Cublington (in Buckinghamshire), Nuthampstead (in northern Hertfordshire) and Foulness (later known as Maplin, on the Essex coast). Stansted was missing as it had already been excluded as being inferior to Nuthampstead on the grounds of air traffic control, of noise impact, and for poorer surface access. Similarly, Foulness had been included as it ranked best on noise, defence, and air traffic control, but it was a very expensive site to develop and was furthest from London with a high surface access cost. Logically, this option should not have been considered further, but it was. The main part of the Commission's work was in the evaluation of the four short-listed sites.

This comprehensive evaluation compared the costs of the sites as compared with the cheapest overall alternative (Cublington). It was argued that the decision was an inter site comparison and the absolute values were meaningless in themselves. Thurleigh was about £70 million more expensive than Cublington whilst Nuthampstead and Foulness were respectively £129 million and £158 million more expensive in terms of their full social costs and benefits. More interestingly, it seemed that these differences were dominated by a relatively few factors. Passenger access costs alone ruled out Foulness, as these costs accounted for £131 million of the difference in total costs. Even when sensitivity analysis was carried out, little change was found in the overall results. The team concluded that even if substantially different values of time were used, the rankings would remain unchanged. Access costs to Foulness far outweighed all other costs, including construction costs. The most thorough application of cost benefit analysis to a major transport infrastructure decision generated tremendous debate over the exclusive use of monetary valuations, for its omission of many planning and economic development factors, for its valuation of certain factors (particularly time and noise), and for its assumption that all people valued money identically (a benefit of £1 was valued identically by people in different income groups). The Commission accepted that the decision should not be based solely on the cost benefit analysis, but that it should be modified by judgement.

Despite the undoubted environmental disadvantages the Roskill team opted for Cublington. Sir Colin Buchanan could not agree with his colleagues and argued that Foulness was the only acceptable site despite the higher costs of development. Public sympathy was with him and, in April 1971, the government announced that the third London airport would be built at Maplin Sands (or Foulness) in Essex.

The Roskill Commission Inquiry decision encapsulates all the main problems inherent in SCBA. The first problem was the forecasting of the level of traffic, which in the case of air relates to the growth in income levels, the value of leisure time, and whether the Channel Tunnel would be built. The Maplin Review (Department of Trade, 1974) scaled down the Roskill predictions on traffic forecasts of air passengers in 1985 for the London area from 82.7 million to a range between 58 and 76 million. The picture is further complicated by the assumptions made on load factors, the size of aircraft and the number of movements possible within the available airspace. It also depended on the number of flights that could be accommodated within the existing London airports (Heathrow, Gatwick, Stansted and Luton). The Maplin Review argued that the determinant factor was not air traffic capacity but terminal capacity, and that the existing four airports could all cope with additional terminals.

In the subsequent debate on the accuracy of forecasts (summarized in Table 3.4), the fundamental argument switched away from a comprehensive evaluation of the alternative sites, as carried out by the Roskill Commission, to a much more restricted assessment of whether the expected growth of traffic could be accommodated within existing airports. In addition it meant that the £650 million (1974 prices) development costs for Maplin could be

Table 3.4 The Maplin Review summary

	1973 Actual (million passengers)	1990 Scenarios to accommodate traffic			
		I	II	III	IV
Heathrow	20.3	38	38	38	53
Gatwick	5.7	16	16	16	25
Stansted	0.2	–	16	4	4
Luton	3.2	–	10	3	3
Maplin	–	28	–	–	–
Provincial airport	–	–	5	24	–
Total	29.4	82[1]	85	85	85

	Capacity increase (passengers/year) (millions)		Airport development costs[2] (£ millions)	Transport capital costs[2] (£ millions)
	from	to		
Heathrow	38	53	115	90
Gatwick	16	25	70	60
Stansted	1	4	15	nil
Luton	4	16	110	47
Maplin	3	10	70	10
Provincial airports	0	28	400	235

Notes
1 Lower as Maplin is assumed to attract fewer passengers
2 Prices at 1974 levels

Source: Department of Trade (1974)

saved. It was therefore not surprising that the decision to build Maplin was abandoned (July 1974). Both the recommendations of the Roskill Commission and the most comprehensive evaluation ever undertaken of a transport investment decision had been rejected in favour of more modest developments at existing airports. In retrospect, this may have been the obvious answer, particularly in light of the uncertainty and global recession created by the oil crisis of 1973–1974.

However, it is ironic that many of the same arguments are again being rehearsed and the Civil Aviation Authority is forecasting a doubling of demand for air travel (1987–2005) with a 13 million shortfall expected in the capacity of the London area airports (Table 3.5). It should also be noted that actual demand in 1987 was 60 million passenger movements, whilst the Roskill research team's estimate (1971) was for 82.7 million passengers in 1985 and the Maplin Review team's estimate (1974) was for 85 million passengers in 1990. The actual growth in demand has been a doubling (1973–1987) whilst Roskill's forecast was a 2.8 times increase (1973–1985) and the Maplin review was for a 2.9 times increase (1973–1990). Forecasting traffic growth has always been problematic both for road and air traffic, and no where is it better illustrated that in this case of air travel.

Table 3.5 Forecast and actual demand for air travel (million passengers)

	1987 Actual	2005 Forecast	1998 Actual	2000 Actual
Heathrow	36	55	60	65
Gatwick	20	30	29	32
Stansted	1	33	9	12
Luton	3	5	5	6
Total	60	123	103	115

Sources: Civil Aviation Authority Forecasts, February 1989, and DETR (2000a) – figures exclude double counting

A second problem is in the valuation of costs and benefits arising from any investment decision. Since all consumers are buying goods and services at the same prices, the equilibrium prices prevailing are a true measure of value. These very strict conditions do not occur, as there are many market imperfections (for example the assumption of perfect knowledge, perfect competition, being able to set prices at marginal cost levels, and the absence of externalities). Necessary avoiding action is taken with the adoption of second best solutions. Yet, in the cost benefit analysis, the exclusive use of monetary values for costs and benefits led to considerable controversy over the use of such methods for evaluation and, despite extensive analysis to support the robustness of the Roskill Commission's recommendations, this nagging doubt still remained. In all calculations, surface access time was dominant and it was here that the value of time for both work and leisure was paramount. If this was the case, then other sites such as Stansted which were close to London should have been included. The irony here was that Nuthampstead was included in the shortlisted sites at the expense of Stansted, but was rejected because of its high noise costs and the loss of agricultural land. Stansted had the same advantages of good access and low construction costs, but less of the disadvantages of noise and agricultural land loss.

Just before the Maplin Review (Department of Trade, 1974), John Heath, who had been Director of the Economic Services Division at the Board of Trade (1964–1970), gave his verdict on the whole debate at a symposium. It is, he says, one of a series of projects with similar characteristics: long delay between the decision and the final implementation, high uncertainty in technological development or in marketing or in both, and large scale. Heath suggests that Britain has made the mistake of going straight for what was thought to be the ultimate solution, without going through the intermediate stages (Hall, 1980). He concluded 'I believe that it is our traditional system of decision making that has let us down in these cases, and that has also landed us with the present controversy over Maplin' (Foster, 1974).

The large-scale public inquiry commission seems to have been a failure and has not lived up to expectations. The facts can seldom settle an issue in themselves and they may lead to restrictive decisions being made at too early a stage. The shortlist of just four sites illustrates this problem. Further, there must be full agreement within the commission itself. Any dissenting view is seen as weakness and allows the decision to be delayed, in order that any perceived challenge to government policy can be dissipated. Roskill encapsulated both of these limitations.

The Leitch Committee's inquiry into trunk road appraisal

The experience with the debate about where to locate London's Third Airport crystallized many of the difficulties in strategic decision-making where a long-term view was needed, where large amounts of public expenditure were at stake, where the cancellation of the decision would be expensive, and where there were important interests vested against any particular decision being made. A similar history could be traced with respect to the roads programme. The public was raising fundamental concerns over road planning, about the forecasts being used, about the valuation of time and accident savings, about the inquiry procedures, and about the need for the actual road. Previously, there had been little debate over the technical expertise of the planner and the boundaries of professionalism had been clearly drawn. However, since the debate over the motorway proposals for London, that expertise was no longer the exclusive right of the professional. The public were both better informed and wanted more information, they were able to employ their own experts and there seemed to be a general feeling in the mid-1970s that there had to be an alternative to building more roads to accommodate the growth in car traffic (Thomson, 1969; Plowden, 1972).

In the 1973 M42 Motorway Inquiry (Coventry to Derby and Nottingham), the Council for the Conservation Society demonstrated that the Inspector could not exclude any objections to the scheme, in particular whether the road should be built at all. Two years later, the Midlands Motorway Action Group tried to present new evidence on traffic forecasts to the M54 Inquiry (Shrewsbury to M6 North of Birmingham). The Road Construction Unit had circulated its forecasts and had stated that this new evidence 'would not be appropriate'. The Inspector ignored the RCU opinion and heard the new evidence. This decision was swiftly followed by a revised Notes for the Guidance of Panel Inspectors (April 1975) which firmly stated the view that traffic forecasts were a matter for government policy and that consultants or DOE officials should not be questioned on them (Levin, 1979). In 1975 a series of major road inquiries were disrupted and in February 1976 the inquiry into the proposed Airedale trunk road was indefinitely adjourned. The average time for a major road proposal to proceed from inception to completion had now increased from 6–7 years in the 1960s to 10–12 years in the 1970s.

It was partly in response to these criticisms of decision-making on road investments, which were now manifesting themselves in a series of high profile demonstrations at major road public inquiries, that the government set up a comprehensive review of trunk road appraisal chaired by Sir George Leitch (Department of Transport, 1977*b*). This high powered review committee was very critical of existing procedures for forecasting, their insensitivity to policy changes, their unintelligibility, and the overriding priority given to road-user benefits at the expense of other factors, particularly environmental concerns and distributional issues. The review covered existing methods and their evolution, the whole procedure in the development of a trunk road scheme, experience from other countries, a series of worked examples and tests, and a thorough review and critique of existing methodology. It presented a framework for the assessment of trunk road schemes. Its most important contribution was to state systematically many of the public's concerns over analysis and forecasting methods, and it vindicated many of the views held by the pressure groups.

As with the air traffic forecasts for London's Third Airport, *road traffic forecasts* were criticized for not taking account of uncertainties, in particular energy and fuel availability and income growth, and for the way in which single figure forecasts were presented without comment on data or modelling problems. It was also suggested that forecasts tend to be self-fulfilling and tend to reinforce existing trends (Adams, 1981).

The *economic appraisal* itself was criticized for being unintelligible to the layperson, inflexible and an impediment to public participation in the decision-making process. The cost benefit analysis used in the appraisal was considered to be partial, with over 85% of the user benefits being accounted for by savings in travel time. Improved safety accounts for a further 19% and vehicle operating costs about 4%. Vehicle operating costs are a negative benefit because road improvements lead to an increase in these costs as a result of higher speeds and longer journeys (National Audit Office, 1988). The value of time is the most important single factor in the evaluation, and the values used for the calculation of work time and leisure time together with the weight given to small savings in travel times are crucial (Hensher, 1989; MVA *et al.*, 1987). The basic question posed by the Leitch Committee, but not answered, was whether people actually save time or merely travel further, and if they do save work time is it transformed into productive use no matter how small that saving. The Committee also expressed concern over the valuation of accident savings and suggested that other factors such as land take, delays during construction, and more robust testing of the capital cost assumptions (often underestimated) should be explicitly included in the evaluation.

Comment was also made on environmental and regional development factors in evaluation, as both of these, although acknowledged as being important, were placed second to road-user benefits. It was accepted that evaluation of the environment was difficult but methods were available to give rankings to these factors (for example the Planning Balance Sheet; Lichfield, 1970). Increased consultation and public presentation would help in explaining options and the implications of each. The recommendations of the Leitch Committee went some way to meeting these criticisms (Department of Transport, 1979a).

In their proposed Comprehensive Framework for the Appraisal of Trunk Road Schemes, the justification for the scheme should include the identification of individual groups affected, a comprehensive and flexible approach that balanced the worth of a project against the extent to which each impact meets a series of standards, regulations and procedures, as is the case in the United States. The assessment should be comprehensible to the public and command their respect. The balance of assessment was switched from its overriding economic base towards a comprehensive framework which relied more on judgement, with full consultation at all stages, and included other factors such as the environment in decision-making.

The Leitch Framework shared many common features with the broader based evaluation procedures developed in planning. In this sense, another strong link was being proposed between transport and planning evaluation. Most similarity was apparent with Lichfield's Planning Balance Sheet Appraisal (PBSA) (Lichfield, 1970; Lichfield et al., 1976) which had been extensively used with the Roskill Commission's Inquiry. Essentially, the PBSA is a modified version of social cost benefit analysis (SCBA) as it deals with monetary valuations where possible, but it differs in that it explicitly differentiates between the various incident groups who gain or lose from a particular investment decision. As such it responds to two problems apparent in SCBA. First, SCBA does not attempt to measure the redistribution of resources and the equity implications of the transactions between different sectors in society. In private investment evaluation and project selection this may not be important, but where broader planning and societal goals are at stake, then it becomes a political consideration. Secondly, although monetary values are used wherever possible, intangibles are assessed without allocating financial values to them. Aggregation takes place where possible, but the Planning Balance Sheet is presented in its entirety. This presentation is important as it allows the decision-maker to see which factors are the most important and some assumptions become more transparent. However, it gives no guidance on how these factors might be incorporated into a decision, other than that impacts on groups might be weighted to account for the equity considerations.

In Hill's Goals Achievement Matrix (Hills, 1973), the evaluation moves from economic measures to a statement of achievable objectives that have been explicitly set prior to the assessment. Each goal is specified and weighted according to its importance, and these goals are set against those sections of the community affected by the consequences of each proposal being assessed. Community incidence groups can be identified by income, social criteria or location, but again each group is weighted for each objective individually or for all objectives combined. Evaluation is by monetary means where possible or by other units, and aggregation is used if the measurement units are compatible.

As with the PBSA and with the Leitch Comprehensive Framework for Appraisal, the whole matrix is presented to the decision-maker as well as the overall interpretation of the evaluation. GAM takes evaluation away from its economic base in the rationality of neo-classical economics and places it firmly in the political arena, where different objectives can be valued according to their importance and where the impacts of decisions on particular groups can be calculated.

Comment on evaluation in the 1970s

These three matrix approaches to evaluation acknowledge the wide range of different impacts in public sector decisions and the fact that many of those impacts are inherently non-comparable. The intention is that the decision-maker can choose a best decision judgementally after reviewing the spectrum of different impacts and making the necessary trade-offs between them. More fundamentally, the Leitch Committee did not address the possible conflicts between the role of the analyst and the decision-maker in public policy. The forecasting process and the analysis of traffic including economic evaluation had all played an important role in demonstrating the demand for new roads. This analysis had been instrumental in gaining Treasury approval and Parliamentary support for the trunk road programme.

However, in practice, as Starkie (1982) comments, 'the combination of long gestation periods for highway schemes and the time taken to develop a standard evaluation procedure meant that economics had had little influence on the shape or sequencing of the trunk road network *constructed* by the early seventies.' The actual influence of COBA and matrix appraisal frameworks on the 2,000 miles of motorway and trunk roads built before the mid-1970s was minimal. Starkie (1982) suggests that it is also debatable whether COBA, even when it was fully operational, had much impact on the shape and size of the more recent roads programme: 'The decision to invest in a road scheme was at the end of the day a matter of judgement alone.' There seems to be a kind of inevitability about the outcomes of various decisions, and analysis may only be appropriate where it supports the preferred outcome.

Perhaps this dilemma is best illustrated with reference to the decision of the Roskill Commission. The first mistake made was the exclusion of Stansted from the shortlisted sites on the grounds that it was inferior to Nuthampstead.

Stansted had been the favoured location of the London airports lobby (consisting of the airlines, the Ministry of Aviation, later the Civil Aviation Authority, and the British Airports Authority) for the previous 20 years, and it was this group that was instrumental in obtaining the Maplin Review. The second error was to allow Buchanan to submit a minority view in favour of Maplin when the rest of the Commission supported Cublington. Any recommendation would have had more weight if it were supported by all the members of the Commission. Also of importance was Buchanan's rejection of the basic economic arguments used by the Commission in favour of a broader planning-based judgement.

The lessons from both the Roskill Commission's recommendation and from the broader use of economic evaluation in transport decisions are clear. To implement successfully any recommendation, it must be unequivocally supported by all members of the Commission or Committee and it must also have strong political support from Westminster. Any weakness in that united front will allow other interest groups to intervene, and to delay any decision being made. Delay is often equivalent to the abandonment of a decision. Intervention can either come from interest groups within the sector, as with the London airports lobby, or from the general public, as with the protests over the roads programme.

With respect to major or controversial investment decisions in the transport sector, action will only take place if all the relevant parties and the government are in agreement. The strength of the case also needs to coincide with the appropriate timing for action, as unpopular decisions are not likely to be taken immediately prior to a national or local election. The opportunities for action are therefore limited.

Evaluation and the environment in the 1980s and 1990s

The level of debate in the 1980s was generally at a much more subdued level than in the previous two decades, when the cost benefit analysis procedures were being set up, when there was a series of large-scale public inquiries, when the protest movement was most active, and when the procedures were reviewed and modified. The 1980s also marked the end of the first generation of motorway construction which had started with the Preston bypass in 1958 and ended with the completion of the M25 around London in October 1986. Certain changes did, however, take place which epitomize three of the major concerns of the 1980s.

Statutory procedures and private sector involvement

One feature of all decisions has been the excessive time taken from project genesis to completion, and the balance between the time taken for preparation (too short) and the time for gestation (too long). On average, Department of Transport schemes were taking 15 years, whilst local authority schemes took 7 years (House of Commons, 1990). In both cases the actual construction time accounted for 2 years. Public discussions took about 30 months, with planning permission, inquiry and compulsory purchases taking a further 3 years. Delays were taking place in Department of Transport procedures between consultations and the publication of the preferred route, and between the close of inquiry and the putting out of contracts. In their evidence to the Transport Committee (House of Commons, 1990), the Department of Transport claimed that they could reduce the total time taken by 4 years with a halving of the time from entry to programme and public consultation (from 6 to 3 years), and between consultation and publication of the preferred route (from 2 years to 1 year). This move towards greater internal efficiency seems a more appropriate means to streamline procedures than the earlier attempts to remove some of the public consultation stages, so that argument could only take place on mitigating circumstances and compensation (Institute of Civil Engineers, 1983). The public inquiry itself only accounted for about a third of the total time, even in the largest schemes. Other proposals (NEDO, 1984) included the use of investigating magistrates (as in France) rather than Inspectors, where the rather passive role of the Inspector in receiving representations and synthesizing masses of information would be replaced by a more active investigative role at the inquiry. This proposal was not acted upon.

The reduction in time taken for the inquiry process also reduces the extent of blight caused by road proposals to people living in corridors where the road might be located (the safeguarded routes). The compensation rules are very strict with market price being paid for properties that are subject to a compulsory purchase order. Again the experience from France suggests that a premium of 10–20% would help alleviate the disruption caused by a forced move. The compensation procedures in Britain seem to be fairly restrictive both in terms of eligibility and the levels of recompense actually paid. More efficient inquiry procedures and more generous compensation terms would help reduce the direct effects of new transport infrastructure on individuals directly or indirectly affected.

The delay also means that the involvement of the private sector in investment in roads, railways and airports becomes more problematical as the risks are seen to increase with the delays. The window of opportunity for investment often follows the economic and building cycles when finance is available, but if delay takes place then that investment possibility also disappears. One exception has been the first road constructed with private finance in the UK, the Birmingham Northern Relief Road, but even here it has only succeeded through the perseverance of the consortium involved and clear financial returns to the investors.

The Birmingham Northern Relief Road – This new 44 km motorway around the north and east of the West Midlands conurbation will provide an alternative route to the M6 between junctions 4 and 11 (Figure 3.2). The BNRR has been designed, built, financed and operated (DBFO) by Midland Expressway Ltd (MEL) under a 53 year concession agreed in 1992 with tolls being charged on users. It is designed as a three-lane dual carriageway motorway with six tolling stations along the route (Table 3.6).

Construction was due to start in 1996 and the road should have been opened in 1998, but due to delays in the public inquiry and the challenge in the High Court more than 3 years delay has taken place. The costs to the private

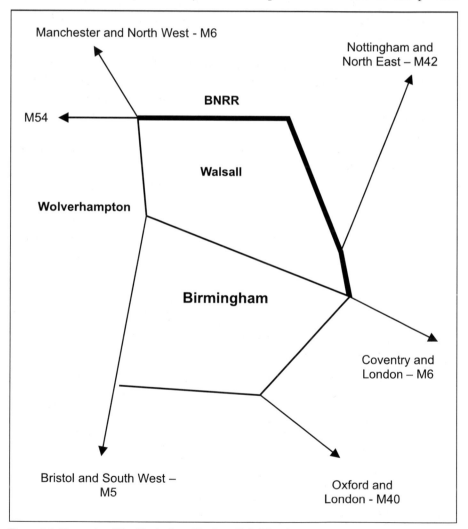

Figure 3.2 The route of the Birmingham Northern Relief Road

consortium (MEL) of the delay and the difficulties of raising the £370m capital in the financial markets might have meant that this road would never have been built.

There were 11,000 objections and representations to the scheme (of which 229 were in support), with 81 alternative schemes being proposed before the inquiries were opened. The main objections to the BNRR were:

- The new tolled motorway was not needed as local roads could be improved to accommodate extra traffic moving through the region.
- The BNRR would not relieve congestion on the M6, but generate more local traffic.
- Concerns were also expressed over increased noise, air pollution, adverse effects on local flora and fauna, and the loss of green belt land.
- The shortcomings of the environment statement.

The main support for the scheme came from the Birmingham Chamber of Commerce, the East Mercia Chamber of Commerce and Industry, the North West Region of the Confederation of British Industry, the British Road Federation, the Freight Transport Association and the Transport and General Workers Union.

The Inspector concluded that the need for the BNRR had been established and that the published scheme was the best available, subject to certain modifications. These were relatively small scale and detailed, designed to ameliorate some of the most adverse effects of the new road (for example, planting, sound barriers, air quality monitoring). This view was supported by the Secretary of State in his decision. The decision to proceed with the road has important implications for the UK roads programme and for the

Table 3.6 Summary of the history of the Birmingham Northern Relief Road

1980	Plans for the Birmingham Northern Relief Road (BNRR) announced.
1987	Extensive consultations with draft orders published for a 44 km motorway from M54 at Featherstone to the M6 near Coleshill.
1988	Public Inquiry held.
1989	Government announced a competition to design, build, finance and operate (DBFO) the BNRR as the first purpose built inland tolled motorway in Britain.
1991	Midland Expressway Limited (MEL) – a joint venture between Trafalgar House and Iritecna announced the winner.
1992	Concession Agreement signed for 53 years. Revised preferred route published.
1993	Draft compulsory purchase order authorising the acquisition of land published.
1994	Second Public Inquiry – Inspector Sir John Fitzpatrick.
1996	Report of the Inspector submitted to the Secretary of State.
July 1997	Secretary of State's decision to accept the Inspector's report.
Feb 1998	Scheme can now proceed. Legal challenge in the High Court.
2001	Construction starts and due to open in Spring 2004.

Source: Based on Banister, Gérardin and Viegas (1998)

consultation process. The strategic importance of the road meant that its benefits (accessibility, safety and economy) outweigh its environmental disadvantages. Issues such as the generation of traffic, the impact on public transport and the quality of the environmental statement were dismissed. All but 1.3 km of the BNRR passes through green belt, even though road construction is not considered an appropriate use of green belt (Planning Policy Guidance Note – PPG2), unless there are special circumstances. The decision means that the traffic concerns (the need for the BNRR) and the lack of alternatives for the route far outweighs green belt concerns. This debate has always been at the heart of protesters' concerns and their view that the consultation process is ineffective in getting policy to change. Their view is that traffic arguments always prevail over the other arguments.

As the BNRR is a privately funded project, there are no controls over the levels of tolls that MEL can charge and so they can adopt strategies to maximize profits, including pricing off heavy lorries so that maintenance costs can be reduced. The exact nature of the concession agreement is confidential and only a general statement has been published. In February 1998, there was a legal challenge in the High Court. An alliance of community groups demanded access to the contractual agreements for the BNRR. The challenge questioned the traffic impacts as it claimed that the new road would only give temporary relief, with a massive bottleneck being created where the BNRR rejoins the M6 in the North, and that a full analysis of its environmental impacts had not been included in the decision. Any means to delay the start of construction were used by local residents and environmental pressure groups. Despite these delays construction has now started and it is expected that the road will be open for traffic in Spring 2004.

More generally, the Highways Agency has the responsibility for delivery of the government's roads programme on time and budget, including the cost effective and efficient management of the 9600 miles of motorways and trunk roads (15,000 km or 4% of the total GB network). As part of its role the Highways Agency has explored the means to reduce the time taken for road building, principally through more open discussion of the options at an early stage before public consultation begins (through a round table conference). The pre-inquiry meetings, where the main issues for discussion are agreed with all parties and a timetable is set up, have taken on a more formal role. Statements of evidence are exchanged between parties before the inquiry and a stricter timetable for the production of the Inspector's report have been introduced. Procedures are now being taken forward simultaneously (rather than sequentially) if there are only minor modifications required, so the delaying tactics can be reduced. Yet the public inquiry element of the process is normally fairly short, with most inquiries lasting less than 16 days. It is the lack of funding and other sources of delay (the government decision process and contracting) which cause most of the time loss in road construction (DOE, 1996 – Circular 5/96).

Use of Parliamentary procedures

A second issue has been the increased use of Parliamentary procedures to speed up the planning process. It is agreed that major developments involving public resources should be assessed and justified in public to determine whether the project is needed, to assess the opportunity costs and the environmental impact, and to compare the selected alternative with other possibilities. But is the same process required when private resources are being used, where the risk is to the developer, where confidence has to be maintained by investors, and where long lead times would reduce the interest of the private sector in the project? The Select Committee procedure has been used for approving major transport investment projects such as the Channel Tunnel and the M25 Bridge which duplicates the Dartford Tunnels. The Commons and Lords Select Committees operate to a very strict timetable and have the responsibility for allocating construction and operating rights as well as land acquisition powers to the developer. Public hearings are only available for those directly affected, and there is no discussion of the wider impacts of the scheme, nor is there a requirement for the major parties to present their case. There is no debate over the alternatives, only the actual scheme, and even then it is not a question of whether it will be built but of minor changes to the proposal. The final decision lies with Parliament.

The Channel Tunnel Bill was a hybrid bill sponsored by the Department of Transport, Eurotunnel, British Rail and Kent County Council. The Commons Select Committee had seven members and its remit was to resolve the problems raised by the groups submitting evidence, but not to challenge the basic proposal in the Bill. In the House of Lords there was only one meeting of the Select Committee which concentrated on the most important problems identified by the Commons Select Committee. The whole review process took under one year, the Bill received the Royal Assent in July 1987, and the Channel Tunnel was opened for public service in 1994. This has proved a faster procedure. The Tunnel now accounts for about a half of all cross-channel traffic (Ross, 1998). These hybrid bill procedures have now been replaced in part by Special Development Orders (since 1990) where Parliament approves the broad principle of a project, leaving matters of detail and local issues to be examined at a relatively short local public inquiry.

Environmental evaluation

The third issue has been the environmental effects of proposed road schemes and their assessment. The Standing Advisory Committee on Trunk Road

Assessment turned its attention to urban road appraisal, and the measurement and evaluation of the environment (Department of Transport, 1986*a*). Their recommendations were that road appraisal in urban areas should be integrated with the wider land-use and environmental impacts, and that consultation with the public and authorities directly affected in the proposals should take place at an early stage. Each project should have a clear statement of objectives together with the relevant transport problems and a range of feasible options which would satisfy those objectives. Appraisal should cover traffic, economic evaluation and an assessment of environmental and social impacts. The basic framework proposed by Leitch (Department of Transport, 1977*b*) had been extended to include a wider range of impacts together with an explicit declaration on the objectives of each scheme, which in turn reflected a move towards Goals Achievement Matrix analysis. Evaluation had moved away from single-criteria analysis to multi-criteria analysis. Although the government's response was not fully supportive, many of the recommendations were accepted (Department of Transport, 1986*b*).

Since the report of the Jefferson Committee to the Departments of the Environment and Transport (Department of Transport, 1976*b*), a restricted interpretation of the environment has been included in the assessment. Environmental costs have been internalized within the decision-making process and only local factors have been considered. The broader effects of transport on pollution, the use of resources, and development are largely ignored (Table 3.7). The *Manual of Environmental Appraisal* (Department of Transport, 1983) develops a list of environmental factors and adds to it assessment of heritage and conservation areas as well as ecological factors.

In June 1985, the EC issued the Directive on the Assessment of the Effects of Certain Public and Private Projects on the Environment, including the construction of motorways and express roads – interpreted in the UK as trunk roads over 10 km in length (Article 4(1)) – and other roads or urban development projects at the discretion of individual Member countries (Article 4(2)). The Directive took effect from July 1988 (EC Council Directive, 1985). The SACTRA Report (Department of Transport, 1986*a*) concluded that their recommendations were compatible with the requirements of the Directive. However, the Directive is explicit in specifying a much wider range of factors that should be included in any Environmental Impact Assessment (EIA). Article 3 requires an assessment of the direct and indirect effects of a project covering the following factors:

+ human beings, fauna and flora;
+ soil, water, air, climate and landscape;
+ the interaction between the factors mentioned above;
+ material assets and the cultural heritage.

Table 3.7 Environmental impacts of transport in the United Kingdom

Environmental categories	Environmental impacts	Transport's contribution (1995 unless otherwise stated)
Energy and mineral resources	➢ Energy resources used for transport (mainly oil-based) ➢ Extraction of infrastructure construction materials	➢ 44.8 million tonnes of petroleum consumed by transport ➢ transport accounts for approximately one-third of the UK's total energy consumption ➢ approximately 120,000 tonnes of aggregates per kilometre of 3-lane motorway ➢ 78 million tonnes of roadstone extracted
Land resources	➢ Land used for infrastructure	➢ approximately 4.2 hectares of land per kilometre of 3-lane motorway ➢ 1,725 hectares of rural land developed for transport and utilities per annum (1992)
Water resources	➢ Surface and groundwater pollution by surface run-off ➢ Changes to water systems by infrastructure construction ➢ Pollution from oil spillage	➢ 25% of water pollution incidents in England and Wales caused by oil ➢ 585 oil spills reported in the UK ➢ 142 oil spills requiring clean up in the UK
Air quality	➢ Global pollutants (such as carbon dioxide) ➢ Local pollutants (such as carbon monoxide, nitrogen oxides, particulate matter, volatile organic compounds)	➢ 25% of the UK's carbon dioxide emissions (CO_2) ➢ 76% of the UK's emissions of carbon monoxide (CO) ➢ 56% of the UK's emissions of nitrogen oxides (NO_X) ➢ 51% of the UK's emissions of black smoke (particulates) ➢ 40% of UK emissions of volatile organic compounds (VOCs)
Solid waste	➢ Scrapped vehicles ➢ Waste oil and tyres	➢ approximately 1.5 million vehicles scrapped annually ➢ more than 40 million scrap tyres annually
Biodiversity	➢ Partition or destruction of wildlife habitats from infrastructure construction	
Noise and vibration	➢ Noise and vibration near main roads, railway lines and airports	➢ approximately 3,500 complaints about noise from road traffic ➢ approximately 6,500 complaints about noise from air traffic
Built environment	➢ Structural damage to infrastructure (e.g. road surfaces, bridges) ➢ Property damage from accidents ➢ Building corrosion from local pollutants	➢ more than £15 million annual road damage costs
Health	➢ Deaths and injuries from accidents ➢ Noise disturbance ➢ Illness and premature death from local pollutants	➢ 3,500 deaths ➢ 44,000 serious injuries ➢ 49% of people who can hear noise from aircraft or trains consider it a nuisance (1991) ➢ 63% of people who can hear noise from road traffic consider it a nuisance (1991) ➢ between 12,000 and 24,000 premature deaths due to air pollution ➢ between 14,000 and 24,000 hospital admissions and re-admissions may be associated with air pollution

Sources: Banister (1998); Central Statistical Office (1997); Committee on the Medical Effects of Air Pollutants (1998); Department of the Environment, Transport and the Regions (1997d, e, f and g); Department of Trade and Industry (1997); Maddison et al. (1996); OECD (1988) and Royal Commission on Environmental Pollution (1994)

Article 5 requires the developer to supply information on the following topics and to have regard to current knowledge and methods of assessment:

◆   a description of the project comprising information on the site, design and size of the project;
◆   a description of the measures envisaged in order to avoid, reduce and, if possible, remedy significant adverse effects;
◆   the data required to identify and assess the main effects which the project is likely to have on the environment;
◆   a non-technical summary of the above information.

The EC Directive provides clear guidance on how a comprehensive Environmental Impact Assessment (EIA) can be prepared. EIA refers to the evaluation of the environmental improvement likely to arise from a major project significantly affecting the environment. Consultation and participation are integral to this evaluation, the results of which must be available in the form of an Environmental Impact Statement (EIS), before a decision is given on whether or not a project should proceed (Wood and Jones, 1991). Additional responsibilities have also been placed on promoters of schemes through environmental legislation relating to EIA (noted above, in EEC Directive 85/337 effective from 1988), air quality management (1995) and new targets for reductions in road traffic (1998).

The *Manual of Environmental Appraisal* meets the requirements of the EC Directive and public inquiries are held for all major schemes to which there are objections, and the environmental assessment will be considered at that inquiry. There is an enormous literature on EIA in the United States where there has been a requirement since 1969 (National Environmental Policy Act) for Federal agencies and those in receipt of Federal funds to prepare EISs for 'major actions'. Such actions include any alterations in land use, planned growth, travel patterns or natural or man made resources. All new freeways, expressways or belt highways, or reconstructed arterial highways providing substantially improved access to an area are included within the NEPA requirements.

In the UK, the Town and Country Planning regulations (July 1988) make it compulsory for some types of project to have an EIA to accompany the planning application. This requirement is in line with the EC Directive and ensures the EIA's compatibility with existing planning procedures. Schedule One includes major transport infrastructure proposals and aerodromes with runways over 2100 metres as projects requiring an EIA. If the proposed development is of sufficient size or is in an environmentally sensitive area the local authority may require an EIS – a summary document. Once the EIS has been submitted, the normal planning procedures are followed with the exception that the local authority has 16 weeks to make a decision rather than the normal 8 weeks. The SACTRA Report on the environmental effects of proposed road schemes and their assessment (Department of Transport, 1986*b*) commented on the full implementation of the EC Directive and on the possibilities for the valuation of environmental factors in assessment.

As the requirement for a full EIA is now part of EC policy and as it has been included within the planning process, it would seem logical to extend explicitly the requirement to the assessment of all new road schemes in the UK. EIA allows the broader environmental issues to be fully presented without lessening the importance of economic factors in assessment. A full EIA is compulsory in France and it also forms an integral part of the road planning process in West Germany.

There seems to be a strong case for the extension of existing economic-based evaluation methods to encompass a broader range of environmental, social and distributional factors. The SACTRA Report goes some way towards arguing the case for such a change (Wood, 1992), but stops short of a full commitment because of many unresolved issues, such as global and local impacts, the complexity of the environment, valuation of the environment, and the general unreliability of forecasts. Yet decisions still have to be made and many controversial issues (for example Twyford Down near Winchester (1992) and Oxleas Wood in London (1993)) have been resolved without a full environmental appraisal. This in turn has resulted in confrontation between European and national government, and has provided environmental groups with a new impetus for action.

Many changes in evaluation have taken place recently (see Chapter 6). The theoretical debate has concentrated on whether the transport benefits reflect the total economic benefits of any scheme (SACTRA, 1999). This may be true if the assumption of perfect competition holds, but in many cases it is now realized that the calculation of transport costs and benefits will not give the full economic impact. It is here that environmental costs represent real economic resources, even when monetary values cannot be calculated. Similarly, real economies are not perfectly competitive, so that some wider economic impacts will not be measured. SACTRA concluded (1999, para 34) that

> these additional economic impacts, over and above the value of direct transport impacts, may be either positive or negative, depending on whether prices are higher or lower than marginal social costs, which in turn depends on the combined effect of divergences between price and marginal cost of output, taxes and subsidies, and unchanged external costs. Therefore there will be some conditions where including wider economic impacts would lead to an increase in the value for money of a transport investment, compared with a conventional appraisal, and other conditions where including these wider impacts would lead to a reduction in the value for money.

In addition to the environmental costs and the imperfections of the economy, there are the spatial distribution effects. There needs to be some assessment of whether the economic impacts are properly targeted at the intended location, or whether other locations (for example, the centre) benefit. Distributional analysis should encompass both winners and losers (in terms of people, businesses and locations), and this in part suggests a framework approach such as that developed by Lichfield (1996) and others, as well as including a microeconomic analysis of economic and other effects (Banister and Berechman, 2000).

Some of these unresolved questions are tackled through the objectives-led approaches which maximize a set of socially (or spatially) based objectives rather than market values. Multicriteria analysis (MCA) offers one such approach, which is

> based on the objectives of the responsible decision-makers, a group of impacts is defined which between them capture the performance level of each alternative project in achieving the set objectives. Unlike CBA, achievement of objectives can be assessed in a number of ways, such as measured quantity, qualitative assessment or rating. These assessments are then transformed onto a scale (typically 0-100) giving a score for each impact for each project. The overall performance of the project can then be estimated by producing an overall project score, calculated by multiplying each impact score by a relative weight for that impact (reflecting the importance with respect to the other impacts) and then summing over all impacts. As with the CBA, there are many potential complexities and issues involved, including the identification and definition of the impacts to be included, specification of the measurement methods and scoring of each impact, the issues surrounding the use of weights, and the means by which scores and weights can be combined to give an overall project score. (Grant-Muller et al., 2001, p. 240)

MCA should be seen as complementary to CBA and EIA, but care needs to be exercised when combining aspects of all three methods, as there is a risk of double counting. The key element is in presenting all the information to the decision-maker in a clear and unambiguous manner. It is a classic example of the dilemma facing all analysis, where the complexity of the reality has to be simplified to help decision-making, but a balance must be struck between methods that achieve a high technical merit and those that achieve a high level of transparency.

Comments on evaluation in the 1980s and 1990s

The questions of public involvement and acceptability have a long history, with the principles of information, consultation and participation being clearly established as the means to achieve this. Yet, the practice still leaves many unanswered problems. The government made a policy statement on modernizing planning (DETR, 1998c).

- A significant European Dimension will be introduced into the planning system through the European Spatial Development Perspective (ESDP) and the INTERREG IIC spatial planning initiative (which promotes transnational co-operation on land-use planning). But most decisions will still be taken locally.

- National policy objectives, particularly on major infrastructure projects, will be given through national policy guidance including site selection criteria. Public inquiry procedures will be streamlined through the setting of dates for the Inspector's report and new duties to stick to agreed timetables. Greater use will be made of Parliamentary procedures to establish approval in principle of a project, but leaving matters of detail to a short local public inquiry. In the last 15 years, there have only been about 10 projects which were national in scale and required an inquiry of more than 3 months.

- Regional planning will have an enhanced role through broad-based Regional Planning Guidance which will set out regional strategy to be discussed at a non-statutory examination in public. The regional strategy will cover environmental appraisal and include an integrated transport policy to support the regional economic priorities of the Regional Development Agencies. This has been summarized in PPG11 on Regional Planning.

- Local efficiency has to be enhanced as 95% of all planning applications are dealt with by local authorities. Instrumental in this plan-led process are the development plans which provide the framework for each local authority to take decisions. This has been completed in the from of the revised Planning Policy Guidance (PPGs) for development (PPG12) and transport (PPG13).

As part of this process, particular attention has been given to the problems created by major infrastructure projects, as these decisions need to be taken at the regional level. Many changes have taken place in the involvement of the public in decision-making and the procedures used in the UK are thorough and fair. The new proposals further clarify the different levels at which decisions should be made, giving a new importance to the European and regional dimensions. Although there is still much discussion about particular cases (for example, London's Third Airport 1968–1971) and new *causes célèbres* (for example, the Heathrow Terminal 5 Inquiry which lasted for 5 years – Banister and Berechman, 2000), most inquiries are short and well managed. In the whole period of construction (10–13 years), the inquiry part is often less than 3 months, including the preparation of the Inspector's report. It is the other delays, principally caused by the lack of finance, the deliberations over road policy, and the detailed procedures relating to land acquisition that result in greater problems. The requirements of the private sector relate to different financing packages and their own shorter-term building cycles, and this new complexity has also caused additional difficulties and delays. The whole of the urban road infrastructure programme in the UK has now been put on 'hold', and there is an intense debate over whether any new roads should be built, particularly if the preliminary evidence on the traffic generation effects of roads is supported and the environmental agenda continues to predominate. But this reassessment may only be short as the government already has new plans for road building presented as part of their 10 year strategy (DETR, 2000*b*).

Chapter 4

Radical policy change

The market alternative to transport provision – the Conservative years

Since the 1960s governments have always played a major interventionist role in transport decisions. The underlying philosophy was that transport should be made available to meet the needs of the population and that everyone could expect a minimum level of mobility, almost as a right. To this end, once the basic motorway and trunk road networks had been established, investment was switched to public transport in the form of capital and revenue expenditure. In the 1980s, the role of government was significantly reduced and market forces were allowed to determine both the quantity and to a great extent the quality of public transport services. Government policy on transport maintained that services should wherever possible be provided by the private sector, services should be determined competitively, not in a coordinated fashion, and fares should be market priced. Coupled with these changes was a move towards greater precision in defining the objectives of public transport enterprises, particularly financial performance objectives and quality of service standards. In practice the reduced role of government was more apparent than real as it could be argued that central government had become more powerful.

It was at the local level that the impacts were really felt, with the abolition of the Greater London Council and the Metropolitan County Councils, the protected expenditure levels, and the deregulation legislation. Similarly, where intervention did take place, it was targeted towards individual initiatives (for example Urban Development Corporations) to correct perceived market distortions.

In transport terms the 1980s was the decade of the motorist with the costs of driving being significantly reduced and the growth in company financing of motoring becoming one of the major private subsidies. In addition to the phenomenal growth in car traffic of some 40% over the decade, there was a parallel increase in overseas travel, principally by air on business and holidays. Over 35% of the population took two or more holidays, and expenditure on leisure accounted for 17% of all household expenditure. The numbers of overseas holidays taken by UK residents trebled from 7 million to 21 million in the period 1976–1988. Society was in transition from one based on work and industry to one in which leisure pursuits dominate – the post-industrial society. This society was highly mobile and increasingly dependent on the car and technology.

The decade was one of optimism, at least until the events of 'Black Monday' in October 1987, with unprecedented growth in house prices, incomes and quality of life for those in employment and not on fixed incomes. It is difficult to conclude whether this wealth was policy induced or whether it was a longer-term cycle of growth in the world economy. Nevertheless, it was a decade of almost continuous activity which has brought about the most fundamental changes in transport policy seen in Britain this century.

- The planning framework within which transport policy was embedded was dismantled with the intermediate tier of local government being abolished. The government dealt directly with the districts (and Boroughs in London) and the statutory transport undertakings (for example British Rail and London Transport).
- The provision of bus services outside London was fully deregulated with services being provided wherever possible by the market at the appropriate price. Subsidy was specific to routes where need was identified and where services could not be provided by the market. In London a system of competitive regulation was introduced (Banister, 1992*b*).
- Transport enterprises were sold to employees, to management, to other companies and to the general public. Most transport related enterprises have now been denationalized, with only the underground system in London still remaining in the public sector.
- A series of measures were introduced to improve safety, such as compulsory wearing of seat belt, changes in the motorcycle test, and changes in the drink-drive regulations.
- Traffic in towns and congestion were identified as key problems and some remedial measures were introduced, such as wheel clamps and the towing away of illegally parked vehicles, 'Red Routes' in London, and increases in parking fines with enforcement being carried out locally by the district authority or by private contractors.
- The environment was seen as the next major political issue, but direct action in the transport sector was restricted to a reduction in tax levels on unleaded petrol.

The main actions of the Conservative governments of the 1980s and 1990s are summarized in Table 4.1. Underlying all these changes has been one consistent objective, namely to reduce the levels of public expenditure in transport, particularly on the revenue side. Although many of the policies have been justified on other criteria, such as ideology (for example, deregulation), democracy (for example, widening share ownership), efficiency

Table 4.1 The Conservative governments 1979–1997

1979–83	**May: Conservatives elected – majority 43**
	Margaret Thatcher – Prime Minister
	Crude oil $18–24 per barrel
	Increase in public transport fares 15–25%
	Fares Fair campaign in London and other cities
	Armitage Report on Lorries, People and the Environment
	Transport Act 1980 – Deregulation Part I
	Transport Act 1981
Falklands War 1982	Transport Act 1982
	Collapse of Laker 1982
	Seat belt wearing mandatory – January 1983
	Wheel clamps and travel cards in London – May 1983
1983–87	**June: Conservatives re-elected – majority 150**
	Transport Act 1983 – Protected Expenditure Levels
Miners Strike	London Regional Transport Act 1984
March 1984	Local Government Act 1985 – Abolition of GLC and MCCs
National Docks Strike	Capital cards in London – January 1985
	Transport Act 1985 Deregulation Part II
End of Miners Strike	Anglo-French Channel fixed link – February 1986
March 1985	Terminal 4 opened at Heathrow and Piccadilly extension
	Crude oil $10 per barrel
	Completion of M25 – October 1986
	British Airways privatized – January 1987
	Rolls Royce privatized – May 1987
	British Airports Authority privatized – June 1987
1987–92	**June: Conservatives re-elected – majority 105**
	Docklands Light Railway opened – July 1987
Collapse of	City Airport opened – October 1987
Stock Exchange	Merger BA/British Caledonian – November 1987
and gales	National Bus Company privatization – April 1988
October 1987	Central London Rail Study – January 1989
	Roads Programme expanded – May 1989
	New Roads by New Means – May 1989
	Auto guide Electronic Route Guidance piloted – July 1989
	Roads for the Future – February 1990
Iraqi Invasion	London Assessment Study proposals axed – March 1990
of Kuwait	Crude oil rises to $26 per barrel
August 1990	Intercity 225 services between London and Leeds – October 1990
	Major takes over as Prime Minister – November 1990
February 1991	New terminal at Stansted Airport – March 1991
– severe weather	Buses in London to be deregulated – March 1991
	Queen Elizabeth Bridge opened on River Thames – October 1991
1992–1997	**April: Conservatives re-elected – majority 21**
	John Major – Prime Minister
	New post of Minister of Transport for London – April 1992
	Manchester Metrolink started passenger operations – April 1992
	Privatization of British Railways started – July 1992
	Jubilee Line extension construction started – December 1993
	Privatization of motorway service stations – January 1994
	State opening of the Channel Tunnel – May 1994
	Scheduled services between London and Paris/Brussels – November 1994
	5000 mile Sustrans national cycle network supported by government – April 1995
	Fire in Channel Tunnel – November 1996
	Thameslink 2000 Rail project approved – February 1996
	Reorganization of local authorities into Unitary Authorities – April 1996
	Transport policy document 'The Way Forward' published – April 1996

Note
This table is continued as Table 5.5 in the next chapter

(for example, private enterprises are more cost effective than public ones), and accountability (for example, to shareholders), it is the Treasury which has been the prime mover behind the radical transport policies.

Even capital expenditure on renewal and new transport infrastructure was severely constrained by Treasury imposed limits. In the 1980s, replacement expenditure was delayed, and this in turn led to increasing unreliability in the system. As can be seen in Table 4.2, capital expenditure on transport infrastructure remained stable in real terms to 1990, despite increases from 1982 to 1984. It was only in the period 1990–1991 that a substantial increase took place (+50%). The low levels of capital expenditure in the mid-1980s coincided with record growth in the demand for travel by public and private transport, with the annual growth rate in the numbers of new cars peaking at 6% in 1989.

The second feature of the changing pattern of capital expenditure has been the switch from public transport investment to investment in roads. In 1981 public transport accounted for 16% of total capital expenditure, but by 1991 this figure had declined to 5%; in real terms the 1991 figure is about 37% of the 1981 figure. The figures (Table 4.2) reflected a relaxation of public expenditure constraints and new capital schemes for roads, railways and London Transport (Department of Transport, 1989*b*; House of Commons, 1990), but much of the increase in capital expenditure could be seen as merely catching up on previous under-investment in capital renewal. The pattern of

Table 4.2 Central and local government expenditure on transport, Great Britain (£ million)

	1979/80	1982/83	1985/86	1988/89	1990/91	1992/93	1994/95	1996/97
Total	2716	4048	4139	4576	6165	7021	7248	6251
National roads	608	957	1030	1305	2165	2389	2574	2018
Local transport	2108	3091	3111	3271	4001	4632	4675	4233
Capital	770	977	936	892	1323	1764	1677	1284
Current	1338	2114	2173	2379	2677	2868	2998	2949
Concessionary fares	139	248	299	356	430	435	473	480
Revenue support for public transport	271	586	468	347	339	395	535	551
National support for rail[1]	788	1100	1192	748	1553	2951	2023	1874
All expenditure	3851	5195	5495	5904	8338	10692	10342	8715
Real expenditure (1999/2000 prices)	11268	10738	9776	9193	11165	13034	11997	9665

Notes
1 Includes both national rail and London Underground
Total = national roads and local transport
Local transport = capital and current
Current = revenue support, concessionary fares and others
All expenditure = total, rail and other
This table is continued for the Labour government in Table 5.7
Sources: DETR (2000a) and www.wolfbane.com/rpi.htm

expenditure is consistent across the whole of the 17 years of Conservative government, with roads expenditure accounting for about 75% of all public investment in transport. The only 'blip' in the sequence was the preparation for the privatization of the railways, when national support for rail increased by over 50% for the five year period from 1991–92 to 1995–96. This increase does not seem to be at the expense of road investment, but in addition to it.

Increasingly the government was looking to the private sector for capital. The scope for the private sector in urban areas was limited as construction costs were high, development lead times were lengthy, risks were high, and returns were likely to be low unless the investment put the developer in an effective monopoly position. Development rights along the proposed route might make private investment more attractive, as would joint ventures with the government underwriting the risk (Department of Transport, 1989c).

It is on the current expenditure side that the government was most active. In the early 1980s there was a concern over the increasing levels of public support for both bus and rail services with subsidy levels for each amounting to about £1,000 million. The market philosophy was to ensure that all services would be provided at a level and at a price which was determined competitively. Intervention should only take place when the market is seen to fail, for example for social need, but even then the intervention should be specific and clearly identified. This meant that subsidy levels should be minimized and that there should be no cross subsidization between services. Protected expenditure levels were set for each conurbation and these were effectively the maximum permitted subsidy levels. A balance had to be established between the interests of travellers and those ratepayers/poll tax payers and taxpayers who contributed to the subsidy. The net result was a significant increase in fares on all bus and rail services, both prior to and after deregulation (Table 4.3). Subsidy levels were reduced by about 50% in the period immediately following deregulation (Table 4.2), but they have now

Table 4.3 The costs of travel in the UK

	Retail Price Index	Cars	Rail	Bus
1979	47.3	57.4	72.2	42.0
1985	82.2	87.0	79.5	76.3
1989	100	100	100	100
1992	120.2	121.7 (123.5)	128.9	128.8
1995	129.4	133.7 (146.6)	150.4	143.0
1998	141.4	149.6 (177.8)	166.3	158.8
1999	143.6	153.2 (192.8)	172.3	164.5

Notes
Figures in brackets show the increase in fuel prices. Cost reductions for cars have mainly come from the reductions in the purchase costs In 1999 the index of purchase costs = 116.2, maintenance costs = 174.2, tax and insurance costs = 185.3, and fuel costs = 192.8
Source: DETR (2000a) and www.wolfbane.com/rpi.htm

crept back to the levels of the early 1980s (in cash terms). In 1980, there was an approximate balance between government expenditure on capital and current accounts (Table 4.2); the figure for 1991 was 2 to 1 in favour of capital expenditure on a total budget that has remained constant in real terms.

One area of continued concern for government has been the increase in levels of concessionary fares (Table 4.2). Eligibility was based on fairly broad social groups (for example the elderly, school children or disabled), but it is realized that not all individuals within each concessionary group need the same level of concession. More sensitive methods need to be identified to allocate concessionary fares to those in need. The net effect would be a reduction in the total level of subsidy.

If the government is to be consistent in this policy then the subsidy to company cars should also be reduced or eliminated. Over 50% of all new car registrations were in a company name and many organizations also paid for insurance, tax and petrol for their employees. Motorists who do not have to pay the true costs of their travel will use the car more frequently and make more trips. About a third of all vehicles on the road received some form of company assistance. Tax liability was increased over the four years 1987–1990 by 200%, but the benefit was still over £2,000 for a 1200 cc car and over £5,000 for a 2,000 cc car for an individual paying tax at 25% (Reward Group, 1990). Apart from the inconsistencies in subsidy policy between the reduction in subsidy for public transport and the increase in subsidy for the company car, the effectiveness of any policy designed to reduce the non-essential use of the private car (for example, road pricing) is diminished if the user is shielded from paying those additional costs. For the market economy to work effectively, competition in both the public and private transport sectors must be fair.

Transport policy over the 1980s was driven by the desire to reduce levels of public expenditure on transport, particularly on the revenue side, and to reduce the role of government in determining policy. Overall, the government was successful, at least in reducing levels of current expenditure by some 29% in real terms (1981–1991). Since that time, public transport expenditure has increased, principally in the time that John Major was prime minister. Again, that increase may have been due primarily to the privatization of the railways as the main growth was in the early 1990s. By the next election in 1997, the figures had again returned to their 1991 levels. Demand for travel, principally by the private car but also by rail (from 1985 to 1990), continued to increase at an unprecedented rate and road user expenditure on transport contributed over 12% of all Exchequer revenues (some £17 billion in 1989). Over the decade this figure rose by 184% in actual terms (62% in real terms), and this has been one factor in the ability of the government to reduce levels of direct taxation. Since 1990, the growth in travel has tapered off, with only a 5%

increase in the use of cars. One explanation for this change may have been the recession of the early 1990s. There was no real growth in GDP over the 3-year period (1990–1992), and for the remainder of the Conservative administration, GDP growth averaged about 2.7% per annum.

Apart from the increased contribution of transport to Exchequer revenues and the decreases in public expenditure on transport, this sector has been the test bed for the three main policy innovations:

♦ to transfer industries from public sector to private sector ownership: privatization or denationalization;

♦ to introduce competition into public transport services and reduce levels of public subsidy: regulatory reform;

♦ to encourage more private capital in major transport investment projects, particularly in infrastructure.

Privatization

After 1945 there was an extensive programme of nationalization which formed part of Herbert Morrison's plans for the reconstruction of British industry. Almost all of the transport enterprises were placed in public ownership. During its term in office from 1979 onwards, the Conservative governments returned most of these nationalized industries to the private sector, either in their entirety or through partial sales. The reasoning behind this policy was partly ideological and partly financial. The ideological argument was that organizations are more efficient in the private sector, with the normal pressures that competition brings. Public bureaucracies are inefficient suppliers of services because there are no measures of productive efficiency. In the private sector there is a simple measure of performance, namely profitability. The financial argument was that the Exchequer could gain through the sale of companies, and it could save through reductions in the levels of public support paid by means of subsidy or lending to service capital debts.

Giving more freedom to management was crucial as it allowed capital to be raised in a variety of ways on the financial markets. Private sector enterprises were no longer constrained by the external financing limits imposed by government in the early 1980s. Deficits sustained by public transport operations have had an unwelcome effect on the government's public sector borrowing requirement. Privatization also contributed directly to Exchequer revenue and reduced the public sector borrowing requirement as the ratio of government expenditure to national income was reduced. Supply side economics also suggested that inflationary pressures were reduced, tax cuts could be introduced and this in turn would allow more private investment to take place so that productivity, output and profitability were all increased.

This powerful logic was countered by the argument that these gains were short term. The longer-term impacts of selling national assets may result in higher taxation as the government would be foregoing revenue from these nationalized industries. In effect the government had borrowed against future income streams (Rees, 1986).

From Table 4.4 it can be seen that there were two distinct phases of transport privatizations which coincided approximately with the first two Thatcher administrations. During the first phase, the privatizations were small in scale often involving management buyouts (for example, the National Freight Corporation), or a single buyer (for example, for Sealink), or a limited stock exchange flotation (for example, Associated British Ports). The second phase was much larger in scale and reflects the strong ideological commitment

Table 4.4 The privatization of transport in the United Kingdom

National Freight Corporation:	Road haulage operator
February 1982	Sold for £53 million to a consortium of managers, employees and company pensioners. The government paid back £47 million to the Company's pension fund to cover previous underfunding
Associated British Ports:	Ports and property development
February 1983	Part of equity sold
April 1984	Remainder sold by tender offer £34 million raised
British Rail:	Some non-essential assets sold
	British Rail Hotels sold in 1983 for £30 million
July 1984	Sealink Ferries sold to Sea Containers Ltd for £66 million
	Proceeds retained by BR with subsequent adjustments to the borrowing limits
Jaguar:	Luxury car manufacturer which had become a subsidiary of British Leyland
July 1984	Sold for £294 million
British Airways:	One of the leading international airlines.
January 1987	Sold for £892 million
Rolls Royce:	Aeroengine business bought by the government in 1971
May 1987	Sold for £1,360 million
British Airports Authority:	Operates seven of the principal airports in Britain including Heathrow and Gatwick
July 1987	Sold for £l,280 million
National Bus Company:	Consists of 72 subsidiaries which run most bus services outside the main metropolitan and other urban centres
December 1988	Sale completed with gross proceeds of £323 million. Net surplus to government after all debts paid of £89m
British Railways:	Consists of 25 Train Operating Companies (TOCs), Railtrack which owns the infrastructure, the Rail Regulator, the Office of Passenger Rail Franchising, 5 Freight operators, 3 Rolling stock companies and 19 maintenance suppliers
March 1994	Sale completed with total proceeds of £5,300 million, including £2,622 million from the rolling stock companies, nearly £2,000 million from the share flotation and £700 million from other sales (freight and maintenance facilities).

to privatization. It is here, through extensive publicity, that the government encouraged wider share ownership and made provision for the smaller investor to 'buy' part of the newly privatized company, often at an advantageous price. The four transport privatizations in Phase I (1982–84) realized about £500 million and the five sales in Phase II (1987–95) realized over £9,000 million.

The last great privatization was British Railways. This was a manifesto commitment in 1992 and the legislation was rushed through Parliament in 1993 and 1994, with the bulk of the routes being transferred from public sector operations to the private sector in 1995 and 1996. The thinking behind the process was to separate the infrastructure (Railtrack) from the operations of services (TOCs), with passenger services being operated commercially where possible, but with clearly specified minimum levels of service. The operators would in turn lease their trains from the independent rolling stock companies and pay for slots on the rail network. The intention was to provide improved services more closely related to customer requirements, and at the same time to cut costs through productivity improvements and greater cost effectiveness brought about by competition. However, it was not clear how competition could be introduced within the rail industry, as in most cases the franchises gave operators monopoly rights over a particular route.

Two other issues were important in the rail privatization debates. The new companies, including Railtrack, could now borrow money on the financial markets, thus allowing much-needed capital investment to take place in the rail industry. Through organizational (and ownership) changes and through efficiency gains, a revitalization of the railways would take place with increases in patronage, which would in turn result in more investment. The second issue was the role of the railways in an integrated transport strategy, particularly in taking some of the pressure off the congested roads. Privatization was seen by government as being the best means by which a competitive transport system could be established, so that it could play a major role again.

In retrospect, the privatization of the railways has been a disappointment, despite substantial increases in patronage. The industry is too small to operate successfully as a multitude of operators in a competitive market. Many of the franchises have now been amalgamated as a few operators gradually take over (Table 4.5). Even then the operators have difficult problems to address. In 1997–98, the 25 TOCs received £1,800 million in subsidies and raised £2,700 million in revenue, and the plans are to reduce the subsidy to under £1,000 million by 2002–03. The TOCs have to raise revenue and cut costs, but this is difficult as some prices are controlled and as capacity is limited. There is no incentive for Railtrack to increase investment to raise capacity as 97% of its revenues are fixed. The National Audit Office report (1998) on privatization also criticized the sale of the railways and suggested that a phased programme would have doubled the revenues to the government. They were also in favour of a partial rather than a complete privatization.

As a result of these concerns, the Labour government has taken action, not to renationalize the railways, but to place clearer, but more onerous requirements on them. The new Strategic Rail Authority has the responsibility of determining the strategy for the railways and to act as the means by which government investment is channelled to Railtrack. They also set performance targets for the operators and Railtrack, review (and renew) the contracts for the operators, and have responsibility for safety of the rail system. As part of this deal, the government is investing heavily in the rail system through its Ten Year Plan (DETR, 2000b). Some £27 billion will be invested over the next ten years (to 2010), but nearly two-thirds of this will take the form of routine maintenance and there is an expectation that the private sector will jointly fund much of this programme. The railways are in a state of high uncertainty as their future is not clear, either for Railtrack – which was declared insolvent and put into administration in October 2001 – or for the operators. A basic question here is whether the government can allow the operators to declare themselves insolvent if they cannot make a profit as the levels of subsidy falls.

Regulatory reform

Deregulation of the bus industry was the most important policy switch since regulation was first introduced in the 1930s to protect a vibrant growth industry against predatory practices. Deregulation as it affects all urban and

Table 4.5 The private rail operators in 1999

Owner	Operator (passengers millions and route miles)	Comment
National Express Group Rail turnover £755 million Profit £3.6 million	ScotRail (56.7 and 1885 miles) Central (32.4 and 1495 miles) Silverlink (30.7 and 200 miles) Midland Mainline (6.3 and 307 miles) Gatwick Express (3.7 and 27 miles)	Sleeper service not totally reliable, but other services reliable and punctual Sprawling west Midlands commuting network – poor reputation, but improving Commuter services from Birmingham to London – problems of reliability High speed trains from London to Leeds via the East Midlands – good reliability and punctuality Airport railway with good reliability and punctuality – ordering new trains
Compagnie Générale des Eaux Transport Finances not declared	Connex South Eastern (93 and 444 miles) Connex South Central (117.2 and 481 miles)	Commuter services in Kent and Sussex – to phase out all slam door trains, but delays and overcrowding Commuter services in Surrey and Sussex – refused permission to extend franchise – some improvements in service quality

Table 4.5 The private rail operators in 1999 – continued

Stagecoach Rail turnover £314 million Profit £21 million	South West Trains (118.2 and 584 miles) Island Line (0.7 and 8 miles)	Narrowly avoided a £1 million fine for cancelling services after making too many drivers redundant – punctuality poor, but reliability has improved Isle of Wight – run efficiently and to time
Firstgroup Rail turnover £173 million Profit £9.3 million	North Western (27 and 1124 miles) Great Western (16.4 and 850 miles) Great Eastern (51.6 and 164 miles)	Passenger discounts for poor service – reprimanded for increase in delays Late running services, blamed on maintenance. More than 16% of trains late and reliability is falling Commuter services to Essex and Suffolk –disastrous performance in Winter 1998 with only 60% of trains on time
Virgin Rail turnover £500 million Profit £13.5 million	West Coast (15 and 676 miles) Cross Country (12.5 and 1657 miles)	At one point over 50% of trains late, but improved to 15%. New 140 mph (225 km/h) tilting trains to cut journey time from London to Glasgow by 90 minutes in 2005 Sprawling network of routes – liable to delays and unreliability of trains
Prism Rail Rail turnover £455 million Profit £11 million	WAGN (52.8 and 257 miles) Wales and West (13.6 and 1568 miles) Cardiff Railway (6.1 and 86 miles) C2C, formerly LTS (23.7 and 80 miles)	East Anglia to London – performance improving fares cut to get passengers to switch from GNER Immense network linking Brighton, Penzance, Birmingham and Fishguard – performance very variable South Wales valleys – with a poor reputation for reliability Vastly improved services from Essex to London – £16 million investment and new management
MTL Trust Holdings Rail turnover £400 million Profit £4 million	Northern Spirit (42 and 1277 miles) Merseyside Electrics (23 and 75 miles)	Runs services in the North-East – shortage of drivers has led to cancellations – performance indifferent Walkouts by drivers – but recent improvements in all aspects of operations
Sea Containers Finances not separated	GNER (13.7 and 920 miles)	Inherited the best trains and the most recently upgraded lines – improved reliability. Bidding to have franchise extended if investment is made in tilting trains
Go-Ahead Rail Turnover £215 million Profit £11 million	Thameslink (30.2 and 141 miles) Thames Trains (28.5 and 363 miles)	High level of punctuality and reliability – runs trains through London from Brighton to Bedford Poor punctuality – fined £800,000. Delays increased from 8% to 15% triggering discounts
M40 Trains Rail turnover £50.5 million Profit £2 million	Chiltern Railways (8.8 and 163 miles)	Birmingham to London services – had to repay passengers £2.5 million after slump in performance, mainly due to engineering works
GB Railways	Anglia Railways (6 and 348 miles)	Runs fast trains to East Anglia – punctuality falling and discounts triggered, but recently quality has improved

Sources: Annual reports, the Guardian 7 October 1999 and the Independent 29 November 1999

non-urban areas outside London was introduced in two stages. The 1980 Transport Act deregulated long-distance bus services (those with a minimum journey length of 48 kilometres), excluded small vehicles (less than eight seats) from licensing requirements provided that they were not being run as a business, and introduced the idea of trial areas where there would be no requirement for road service licences. The 1985 Transport Act effectively created a trial area for the whole country except London. The commercial network only has to be registered, and the local authority has no control over the operator provided that he or she runs the services as registered. The subsidized network complements the commercial network with each service being put out to tender by the local authority who has to specify the route, levels of service and fares (in most cases). London has been given an intermediate position under the terms of the London Regional Transport Act (1984) with only limited competition through competitive regulation (Banister, 1985; Glaister, 1990).

This two-tier system treats each route on its own merits and there is no commitment to a network of services. The government argued that competition will bring greater efficiency into the public transport industry and an end to the extensive practice of cross subsidization. The main impetus to change had been the concern over the escalating costs of grants and subsidies to the bus industry, which totalled over £1,000 million in 1983.

The deregulation of bus services under both full deregulation (outside London) and competitive regulation (in London) led to significant cost savings per vehicle-kilometre, amounting to some 40% in the metropolitan areas and 20% in London (Table 4.6). It should be noted that although these figures are in real terms, they only relate to the supply side. Costs in terms of pence per passenger journey increased by 24% in the metropolitan areas and by 13% in London (Department of Transport, 1991b). These savings in operating costs include reductions in wage levels as well as improvements in productivity. Savings have been achieved through direct reductions in basic wages, lower wage levels for new staff, and lower wage rates for minibus drivers. The local authorities also provided generous voluntary redundancy schemes which enabled the passenger transport companies to trim staff levels, particularly in engineering and maintenance staff (−41%), but also in platform staff (−7%). In London there has been less scope for reductions in staff costs as salaries in the capital have been at higher levels historically, and there have always been problems in recruitment. In addition to the downward pressures in wages, there have been significant improvements in productivity with increases in shift working, greater time spent actually in service, and greater use of the available vehicles, in particular with the introduction of minibuses. The crucial long-term question is whether this level of savings in the costs of providing the services can be maintained, as most of the savings were made in

Table 4.6 Summary of changes over the first five years of deregulation in London and the metropolitan counties

	Full deregulation: metropolitan counties	Competitive regulation: London
Passenger journeys	−26.2%	+3.9%
Vehicle-kilometres	+12.9%	+11.4%
Fares[1]	+31.8%	+12.0%
Proportion tendered	14%	40%
Stability	Improving	High
Staff employed	−26.0%	−23.0%
Passenger receipts[1]	−12.2%	−9.1%
Revenue support[1]	−61.3%	−18.0%
Concessionary fares[1]	+19.3%	+8.3%
Operating costs[1]	−40%	−20%

Note:
1 Changes denoted in real terms from 1985 to 1991
   The metropolitan counties include all major British cities – West Midlands, Greater Manchester, Merseyside, West Yorkshire, South Yorkshire, Tyne and Wear, and Strathclyde
Source: Department of Transport (1991b)

the first three years of deregulation. Evidence since then does suggest that this is the case and that there are substantial long-term cost savings from bus deregulation

For competition to take place in the bus industry, the market must be contestable. It seems that the conditions of contestability do exist as sunk costs are low and all operators have access to the same technology (Banister, Berechman and De Rus, 1992). The market is less contestable when barriers to entry are considered, such as access to bus stations, the availability of service information and experience, and the requirement for 42 days notice for changes in service. These factors have made it difficult for 'hit and run' operations where the new entrant sees a market opportunity and enters the market for a short while to make a profit and then withdraws when the incumbent operator reacts. The market position of small operators seems to be with the tendered services and not on the commercial network. Competition on the main corridors of the commercial network in the metropolitan areas appears to be restricted to the larger operators where competition takes place among equals and where the incumbent advantages are limited. This means that smaller operators may have a greater potential to compete in London where services are contracted out. The smallest group of bus operators (each covering under 5 million kilometres on local services per annum) increased their share of local markets from 13% to 17% between 1985 and 1987 (Department of Transport, 1991b). Since then, their overall share has remained fairly static, though their importance has increased further in London and the metropolitan areas, principally through successfully competing for the subsidized network.

The natural sequence in the competitive market may be the re-establishment of oligopolistic or monopolistic operations, and it will only be on the tendered services that the small operators will be able to compete as the successful applicant is granted limited monopoly rights through the bidding process. In the metropolitan areas there are now a few large operators running the vast majority of services with small operators running special niche services and some tendered services. It is in London that the greater variety of operators are found as all the competition takes place off the road, and small operators compete with the large operators on equal terms through the tendering process, but even here consolidation of the industry has taken place.

The metropolitan areas have been the crucial test of the 1985 Transport Act and the main effects seem to have been a reduction in revenue support on buses at no great loss to service frequency (Table 4.6). Tyson (1988, 1989, 1990) estimates that about 70% of this saving to the tax and rate payer has been passed onto the traveller through higher fares. However, due to the uncertainty caused by a fairly traumatic period of transition, there has been a significant downturn in patronage which has again had implications on the profitability of bus companies. Permanent damage may have been caused to the quality of bus services.

The net effect of deregulation in the metropolitan areas has been significant increases in fares (+31.8% in real terms between 1985 and 1991) and a decline of 26.2% in passenger journeys. Undoubtedly some radical change was necessary, but competitive regulation or franchising as practised in London may have been more appropriate than full deregulation, as patronage there has been increased (+3.9%) and fares have risen in real terms by 12% (Table 4.6 and Banister and Pickup, 1990). Cost savings can be made through the tendering process and the quality of the service can be maintained through the terms of the contract.

Although significant savings have been obtained through reductions in revenue support, there have been compensating increases in the levels of concessionary fares. During the first five years of deregulation (1986–91), revenue support for bus services outside London decreased by 53% and concessionary fares increased by 10% in real terms. Despite this, the net overall saving was still considerable at nearly 30% (£217 million). If competitive regulation had been introduced in the metropolitan areas then the cost savings achieved in London through the increases in productivity could have been further augmented by the savings made in labour costs due to the differences in the local labour market conditions outside London. Net savings seem to have been similar under the two types of competitive regimes but the costs of full deregulation are much higher as there is much greater instability and loss of patronage. The British experiment has been keenly observed by other European countries but they have not adopted the same model of full deregulation. The competitive regulation option has been adopted in many Scandinavian countries as preferable to full deregulation as it allows control over the levels of service provided through the tendering process, and it also permits some integration through common ticketing and timetables. As with many of the Conservative 'experiments' with the market, it is where the public sector can be combined effectively with the private sector that worthwhile solutions have resulted.

The bus industry has been transformed over the last 15 years, through deregulation and privatization. However, there has also been a consolidation within the industry as the larger companies gain control of the market. Initially, many operators competed for market share through 'flooding' the profitable routes with buses. This period of over supply and 'on the road competition' was relatively short, as alliances and takeovers of the smaller companies by the larger operators followed. At present there are five companies that account for over two-thirds of all buses operated in the UK (Table 4.7). This consolidation reflects the size of the bus market. It is too small (and contracting) to maintain a high level of competition, and the natural tendency is towards monopoly as there are scale economies in the industry. The larger operators have the opportunity to get higher discounts on new buses, they can switch buses more easily between routes, they have the ability to raise capital on the financial markets, and they are seen as being 'cash rich' by investors. Many of the larger operators have also taken the opportunity to diversify into other types of transport operations (for example, rail operations – see Table 4.5) and operations overseas (for example, Stagecoach has interests in Scandinavia (Swebus), in Hong Kong, and is a coach operator in the United States).

Table 4.7 Consolidation of the UK bus industry

1998	First Group	Stagecoach	Arriva	National Express	Go-Ahead
Market share	21.6%	16%	15%	6%	6.4%
Bus turnover	£614m	£492m	£472m	£177m	£201m
Bus profit	£93m	£72m	£50m	£40m	£23m
Share price (pence)					
July 1997	200	650	340	500	450
August 1998	500	1250	420	1100	760
October 2000	230	68[1]	238	905	602
May 2001	296	57	309	915	830
Group turnover	£795m	£1374m	£1421m	£1134m	£423m
Percentage bus	77%	36%	33%	16%	49%

Note
1 Stagecoach had a share dilution in 1999

There now seems to be stability in the bus industry after some 10 years of turmoil following deregulation. With the consolidation of the industry and stability in services, the bus users now have higher frequency services in the major corridors of demand and in the urban areas. In the rural areas, bus services still need to be heavily subsidized as demand levels are low, and it is here that many of the innovations have taken place through the community transport sector and a new generation of minibus services providing flexible routes (Cullinane and Stokes, 1998). There are more buses running now than prior to deregulation, but the costs of using those buses has risen at a rate substantially higher than inflation (Table 4.3), particularly immediately after deregulation (1985) when there was the major period of readjustment. As a result of this and other factors, the demand for buses has continued to fall and deregulation has not redressed this problem.

Infrastructure

Britain completed its first generation of motorway construction with the opening of the last section of the M25 around London in October 1986. Progress has been steady since the building of the Preston Bypass in 1958, and there are now some 2,800 km of motorways. However, it seemed that the 1990s might signal a significant period of new investment, not just in new construction, but also in rebuilding the existing transport network. Much of the urban infrastructure is Victorian in construction and needs extensive renovation. In the transport sector several large new projects have been constructed, including the Channel Tunnel, additional runway and airport capacity in the London region, the electrification of the East Coast main railway, and the widening of several of the first generation of intercity motorways.

Construction of new infrastructure has raised two important dilemmas. The first is the contradiction between the desire for greater mobility and meeting the ever increasing demand for travel, and the concern over the environment. Apart from needing land for new roads (4.2 ha of land is required for each kilometre of motorway), there are also development pressures, conflicts between conservation and landscape objectives and employment and economic objectives, and a wide range of pollution and resource implications, as well as the direct impact of noise, vibration and severance. The relevant factors are complex, and although considerable debate has taken place over environmental concerns, they often only appear as an addition to the evaluation process and are rarely seen as being influential in decision-making. The high-mobility, car-oriented society is incompatible with reduction in levels of pollution, lower consumption of energy, and protection of the environment.

The second dilemma concerns the financing of major new infrastructure investments. In the past, infrastructure has been seen as a public good and funded from the Exchequer. With the reductions in public expenditure, the private sector has been seen as taking more of the risk, putting together the finance, and receiving payment in the form of tolls (or shadow tolls) for the use of the facility. Implementation can be expedited through the use of Parliamentary procedures which avoid the necessity for large, costly and lengthy public inquiries. Examples include the construction of a second bridge across the Severn and the new bridge to supplement the two Dartford Tunnels on the M25.

The Dartford River Crossing Company (Trafalgar House, Kleinwort Benson, Bank of America and Prudential Assurance) has built a suspension bridge to provide four new lanes for southbound traffic on the M25. As part of the deal they have also taken over both existing tunnels which are now used for northbound traffic and they receive all the toll revenue. The new bridge opened in 1991. The problems of finance and public inquiry did not arise and the government took no risk. Although there are controls on the levels of tolls which can be charged, the company has a maximum of 20 years to recoup costs (to 2008) and make a return on investment. As there is no other river crossing within 20 kilometres this return is almost guaranteed and they have achieved their financial objectives in 2001 (DETR, 1999*b*). There will be an additional year of tolls to accommodate a maintenance fund and the crossing will then revert to the government free of debt. There has been an increase of some 70% in traffic across the bridge since it opened (1991–2001).

The most prestigious project is the Channel Tunnel which epitomized the Conservative government's new optimism concerning the future direction of large-scale infrastructure projects, and the use of private capital. The Channel Tunnel Group-France Manche (CTG-FM) received no public funds or financial guarantees, and so it took all the risk itself. The governments gave assurances to investors that there would be no political interference or cancellation, and that the promoter has full commercial freedom to determine policy including fares levels. The Channel Link Treaty was signed in February 1986 and a Private Bill was introduced in Parliament. No public inquiry was held, only Select Committee hearings where a tight schedule allowed only limited discussion of many local concerns, in particular about the site of the tunnel entrance and facilities at Cheriton just north of Folkestone. Construction started on both sides of the Channel in 1987 and the tunnel was officially opened in May 1994, with passenger services between London and Paris/Brussels starting in November of that year. Capital was raised in the City and through a series of public share flotations, with the total costs doubling to nearly £8 billion. Part of the cost overrun was due to the scheme being completed a year later than originally scheduled, and it will be many years before it will make a return for its investors.

The Channel Tunnel provides the procedural model for the approval of large-scale transport investment projects. Consultation procedures, which have often delayed road proposals for up to 10 years and the decision on London's third airport for more than 30 years, have now been short circuited through the use of Parliamentary procedures. The government now takes only limited risks as the capital has been raised privately. The general public may feel that they have been ignored as they have no direct input to decisions that often affect them directly. Private companies are looking for investment opportunities which can guarantee a return or minimize any risk. Projects initiated in this way should fit into an overall transport strategy and these new procedures must not reduce the need for public expenditure on infrastructure. Private sector projects are very much more dependent upon market conditions and assumptions of continued growth in the demand for travel. Consequently, private sector investment should be seen as an addition to the public sector programme, not as a replacement for it.

Through radical changes in policy, the Conservative governments succeeded in reducing the levels of public expenditure on transport (Table 4.2), and this has been offset by proceeds from the privatizations and some private sector involvement in investment. The growth in demand for travel and the increase in car ownership have also resulted in record revenues for government from the transport sector, through taxation and fuel duty increases.

However, even after 17 years of unprecedented activity in the transport sector there has been no statement of transport policy, except that the tenets of a market approach, efficiency, and value for money seem dominant. This dominance is implicit rather than explicit. The question is whether that 'no policy' position is tenable and whether it can be maintained. The nearest the successive Conservative governments came to producing a policy statement on transport was in 1996, when *The Way Forward* (Department of Transport, 1996) was published, but the new Labour government came to power shortly afterwards.

The lessons from the Conservative years

It can be argued that successive governments have avoided a comprehensive statement on transport policy. The Labour government attempted such a commitment in 1976 with *Transport Policy – A Consultation Document* (Department of Transport, 1976a), but their efforts were not implemented as there was a change of government shortly after the Transport Act 1978. The energy crisis, world recession, high inflation and unemployment all proved to be more important issues for government action.

Although the 1980s brought a period of unprecedented activity in transport legislation, there was still no coherent statement of transport policy.

Nevertheless, the 1980s and the 1990s were both 'decades of the car', with traffic increasing by almost 60% and the numbers of cars and taxis by 50%. In real terms the costs of motoring have never been cheaper (Table 4.3), and some two-thirds of UK households had a car (1997). To the user the car offers real advantages which alternative forms of transport can never match except in congested urban areas and over long distances. The car has unique flexibility in that it is always available, it offers door to door transport, and it effectively acts as a detachable extension to the home. This freedom is entirely consistent with the emergence of the ideologies of the new right in the 1980s, which were reinforced by the policy actions.

After a decade of consistent growth in demand for transport, certain major weaknesses and inconsistencies in the policy approaches of the 1980s and 1990s can be identified. Four major concerns will be covered here, and each picks up one of the major switches in policy over the two decades and they provide some of the major new challenges facing analysts and decision-makers at the turn of the century. Some of these issues are returned to in the following two Chapters.

- *Infrastructure investment and replacement* is required, but the practice of relying on the public sector alone and the inclusion of public expenditure on transport infrastructure in the Public Spending Borrowing Requirement all need thorough debate. The appropriate roles of the private and public sectors in transport infrastructure provision need to be established.

- *Congestion* is often seen as the most important problem, but this has been true over the last 30 years as well. Does the new realism actually bring any new thinking?

- Accurate *forecasts* have always been difficult to make, as this anticipates the demand for travel either in a global sense or for particular routes where investment decisions are being made. The demand for new facilities often seems to be overestimated or underestimated, whilst the costs for construction often seem to be underestimated.

- *Regulatory reform and privatization* have brought many short-term benefits, but at a considerable cost in terms of lost patronage. It is unclear whether competition and contestable market conditions really exist in the public transport sector, or whether the market is too small and natural monopolies will result.

In part, some of the answers to these important issues have been covered in this Chapter. However, some of these lessons have already been learned. It is now clear that the private sector cannot replace the public sector for capital investment in the *infrastructure*, and there must be some form of partnership. The responses to the government's Green Paper, *New Roads by New Means*

(Department of Transport, 1989c) were very clear on this. The only projects which the private sector was interested in undertaking were those that might give effective monopoly control (for example, the Dartford Tunnels and Bridge) or those which were necessary to attract tenants to major office developments (for example, the Canary Wharf development funding the Docklands Light Railway extension to Bank). A third possibility might be where development rights would be given as part of the proposed transport investment. An example might be to allow development at a road intersection to be undertaken by the company which had paid for the construction of that road. The reasons for the lack of interest from the private sector even when there was economic growth are clear. There is always the problem of 'free riders'. Why pay for transport infrastructure when it is available at zero cost already?

The case of the Jubilee Line Extension from Charing Cross through the Surrey Docks and the Isle of Dogs to Stratford illustrates the problems. All these areas have benefited from the railway and the rent levels for businesses and private landlords have been considerably enhanced. Yet it was impossible to set up a package where all beneficiaries contributed to part of the construction costs. The largest developer in Docklands (Olympia and York) offered to pay £200 million, but even this contribution was phased so that most was paid after the line was opened. Debate and dispute has led to delay and may have been responsible, at least in part, for the demise of Olympia and York's prime development at Canary Wharf.

The private sector is interested in low risk projects which have a short payback period and provide a good return for their investors. Most transport projects have long pay back periods (20–25 years) with low rates of return. They also involve large capital flows early in the investment, often result in considerable capital overspend, and are expensive to withdraw from once a decision to go ahead has been taken. The history of the Channel Tunnel project also illustrates all the problems of major transport investments (Holliday et al., 1991).

The government recognized the major role that the public sector should play, and investment in road and rail schemes was increased (1989–1991). However, much of this increase was used to catch up with under investment in the early part of the two decades (Table 4.2), and part of it allowed the local authorities to raise the money themselves through credit approvals. A second generation of road construction and upgrading was planned, including 160 bypass schemes, the upgrading of the Al to motorway standard, and the construction of a fourth lane on most of the M25. The National Roads Plan's 1997 investment levels on national roads were some 60% higher in real terms than in 1979 (Department of Transport, 1989b and DETR, 1999b), and the total roads budget for the period 1991–1997 was 28% higher in real terms than over the corresponding period in the previous decade (DETR, 2000a).

Congestion was seen as the major problem facing transport planning and this increased level of investment might help alleviate that congestion. However, it seemed that in the longer term additional capacity might in turn result in more demand, as the levels of cars owned and cars used per kilometre of road increased. This meant that road congestion would get even worse. This problem might be insoluble, but it has raised a series of important questions and it acted to focus attention from all interested parties.

At the planning level it strongly suggested that intervention is required. Despite the heroic efforts of traffic management schemes and more recent technological developments, the transport system has been operating close to capacity for considerable parts of the day. The choice for the government was to decide whether a strategic view conflicted with a free market. Although business had clear commercial objectives concerning its own profitability, it might prefer to operate within a publicly planned environment with a longer-term horizon that reduced uncertainty.

Even though there was no strategy, one has emerged by default. Road building in urban areas did not seem to be strongly supported except to ease particular bottlenecks. Road investment would take place on orbitals and intercity roads, including the widening of motorways and the construction of new toll roads with some private sector involvement. New rail investment was likely to update existing lines through the introduction of new rolling stock and electrification. New underground lines might be constructed in Central London and there were some 20 proposals for light rail transit in other cities, some of which had already been started (for example in Manchester – see Table 4.1). Again, the private sector was being encouraged to make some contribution and to take over the operation of services on the publicly provided infrastructure. Ironically, many of these capital schemes were similar to those promoted in the 1960s when there was a similar period of economic growth.

The new realism (Goodwin *et al.*, 1991) identified the publication of the Department of Transport's *National Road Traffic Forecasts* (Great Britain) (Department of Transport, 1989a) as a watershed. It was forecast that economic growth and existing trends would result in traffic levels increasing by between 83% and 142% from 1988 to 2025. This doubling of traffic meant that whatever road construction policy was adopted, congestion would increase and that the key determinant in policy was now demand management. There was no alternative policy as it was now no longer possible or desirable to build roads to match the expected increase in demand. It now became essential to devise policies which would reduce demand to levels commensurate with available supply through a range and combination of policy levers. Five common themes have been identified:

- improvements in the quality and scale of public transport;
- traffic calming in residential and central locations, and tilting the balance in favour of pedestrians;
- advanced traffic management systems;
- road pricing;
- assessment of the need for new road construction must follow from a consideration of how much traffic it is desirable to provide for, not how much is required to meet demand.

Two basic questions remained unanswered here. One relates to the desire to reduce the total amount of car use and to achieve an appreciable transfer to other modes (Goodwin *et al.*, 1991). All recent evidence suggests that modal switching was in the reverse direction and that public transport had lost much of its patronage to the car. Such a switch back to public transport might not be feasible in capacity terms, let alone achievable in policy terms. More likely would be the continued use of the car, but to more local destinations and even more complex sequencing of trips. This would allow the same activity participation rates but with less travel distance.

The second, closely related issue was that lifestyles had been adopted which were car dependent. Action was also required on the planning front to ensure that location decisions by individuals and companies, together with the availability of facilities, were within closer range of each other. This suggested a much more interventionist planning system that was directive in its approach. Alternatively, the price of land should be raised in locations which were considered to be inaccessible so that development could be 'encouraged' in accessible locations. Accessible in this context meant in close proximity to all people so that travel distances could be minimized. The unresolved issue here was that even if travel was more expensive and trip distances were reduced, what would people choose to do with their free time? In the past, increased leisure time and faster travel had resulted in more remote destinations being selected in preference to local destinations. The 10-minute walk was replaced by a 10-minute drive. Underlying the new realism was the essential need to convince the general public of the benefits in terms of financial costs, time, and social and environmental costs of using local facilities, and in only travelling long distances by car or air if it was essential. In the UK, there was little sign that this attitudinal change was taking place, but in mainland Europe it did seem that people were prepared to accept traffic calmed areas and reduced speeds by car.

Related to the realization that forecast demand cannot be met by new road construction was the concern over the forecasts themselves and the *forecasting process*. These forecasts have been essential for the proper allocation of public resources. This covered the ranking of possible new road schemes and helping to decide on the total budget for roads in competition with other areas of public expenditure. Scheme forecasts were mean estimates of traffic volumes on a particular project under consideration. They were obtained from estimates of existing levels of traffic on the road network, and national or local growth factors were then applied to estimate future levels of traffic. A model was used to allocate traffic to individual routes (National Audit Office, 1988). In some cases additional trips expected from new land-use developments were added to the traffic growth, but other forms of traffic generation and redistribution were not normally included.

In the 1970s the criticism was based on the forecasts being too high and there was too much capacity being designed into the system. Since then the reverse was true with new roads reaching their design capacity almost as soon as they were opened – 'the M25 effect'. Forecasts of national traffic growth over the 1980s had been too low. Part of the explanation is the growth in GDP which took place between 1982 and 1987 (+18%) and the consequent growth in car ownership, reinforced by the reductions in the real costs of petrol, and the impact that these factors have had on car use (Figure 1.1). More fundamental though was the restructuring of industry, the move out of the city, and the changes in lifestyle which have taken place over the last decade. All these changes were facilitated by the car and patterns of activity evolved which were car dependent.

The government claimed that 50% of road schemes were within 20% of actual forecast levels. 'This is a reasonable record given the inherent difficulties in forecasting, and the number of years, typically six or seven, between the time of the forecast and the outturn' (Department of Transport, 1991a, para 3.4.6). There was no pattern as the forecast was just as likely to be an underestimate as an overestimate with an average error of 28% in England (19% in Wales) (National Audit Office, 1988). In the North, the North-West and the West Midlands traffic flows in most schemes were well below those forecast. In the East and the South-West the reverse was true. At the individual scheme level the variation was even more marked. One conclusion could be that the models were being mis-specified with the salient variables being omitted. Another explanation could be the level of generated and distributed traffic associated with new road construction, suggesting that the trip matrix was not fixed. Certainly, an increased attention to sensitivity analysis and the costs of being wrong was essential in helping to ensure that the allocation of public funds to roads was carried out on the best available analysis (House of Commons, 1990). It was also suggested that the degree of weight placed on the latest forecasts (Department of Transport, 1989a) should be modified in light of the inaccuracy in past forecasts and the fact that no significant changes in methodology had taken place. The new realism among transport experts might also have become modified as the forecasts were again proved to be too optimistic or too pessimistic. Similar problems with the accuracy of forecasts of both demand and costs were experienced in Europe and elsewhere (Skamris and Flyvbjerg, 1997).

The experience of *regulatory reform* in the bus industry and the *privatization* of many transport enterprises brought benefits, but it was unclear whether the services provided were more efficient or more competitive. Certainly, productivity increased, subsidies to the service were reduced, and most companies were still in operation. But subsidies to the user had increased (for example, through concessionary fares), fares had increased, and patronage had fallen by a greater rate than trends might suggest.

In a deregulated market, economic efficiency was promoted either through many operators competing within the market or through the maintenance of conditions of contestability, principally through freedom of entry and exit to and from that market. The main objectives of existing operators had been to maintain their monopolistic or oligopolistic control of that market, both through anticipatory action and the erection of barriers to entry including collusion with other operators, and through predatory action when entry did occur (Banister, Berechman and De Rus, 1992).

Natural monopolies if they existed might create barriers, threaten retaliation and violate the assumptions of ultra free entry to the market. Only if real competition was possible would entry actually occur and this would lead to the internal structure of the industry evolving towards competition. High barriers led to high market shares, concentration within the market and higher than normal long-run profits. Conversely, one operator could be given limited monopoly rights to run services on a particular route. Competition had taken place off the road in the tendering process and not on the road through direct competition.

Deregulation helped in reducing barriers to entry and in allowing competition to take place, but there were many strategies that the incumbent operator could pursue to maintain market position. For example, the industry has restructured itself into larger units through mergers and the formation of holding companies, costs have been significantly reduced, and the network has been rationalized with a clearer indication of cost allocation to particular routes. Consolidation within the industry has taken place. Innovation through improved vehicle quality and minibus operations have been some of the main benefits of deregulation in the bus industry. The losses have been the fares re-adjustments and a period of intensive instability, with poor information and confusion for the customer. Subsidy has been reduced but patronage has also declined significantly on urban bus services. It is still unclear whether bus operations are more efficient in the private sector, whether the market is contestable, or whether the bus industry is a natural monopoly.

For other parts of the transport industry, the benefits of regulatory reform and privatization were clearer. Associated British Ports, the British Airports Authority, British Airways, some of British Rail's subsidiaries, the National Freight Corporation, and Jaguar had all benefited from their return to the private sector. Efficiency and productivity improved, with profitability and return on investment being clear measures of performance. Greater autonomy and control have been given to management, with clear corporate objectives and accountability. These arguments may seem attractive, but Kay and Thompson (1986) have argued that the case was less than clear. The policy of privatization did not have a sophisticated rationale, rather it lacked any clear analysis of purposes or effects, hence any objective which seemed achievable was seized as justification. The outcome was that no objectives were effectively attained, and in particular that economic efficiency had been subordinated to other goals, in particular political goals.

This argument was well illustrated by the debate in the 1990s over the privatization of the railways where there was considerable political impetus for action, but where the alternative forms of privatization were not seen to be attractive and were not debated extensively. It was accepted that it would have been very difficult to maintain a competitive market for rail operations on a complete network basis. Alternative forms of privatization could have included:

- Parts of the network could have been operated within the private sector (for example freight).
- The railways could have been operated on a regional basis as before nationalization (1945).
- Track ownership could have been separated from the operation of services, with the track being kept in public ownership and the operation of services in the private sector.

It should be remembered that the railways have always lost money, despite extensive network and service rationalization, and the use of property sales to finance investment. It now seems likely that this position will not improve. Within the transport sector, the railways are a natural monopoly and it is difficult to see how competition can or should be introduced within the sector as a whole (Banister, 1990). If there is a competitive position for the railways, it is with the car and air over the medium to long distance hauls (200 km) and on the commuting services into the major cities. The political arguments dominated both the economic and planning arguments in the 1980s and 1990s.

Chapter 5

Contemporary transport policy

Introduction

The last ten years (to 2002) have seen two significant changes to transport policy. The first is the new environmental debate and the challenge of sustainable development. There has always been a realization that transport has significant environmental costs, but in the past these have mainly related to local environmental issues, such as noise, severance, visual intrusion and some pollutants. The new debate is much broader and includes the global pollutants, acid rain, the use of non-renewable resources, and the health effects of transport (see Table 3.7). In this Chapter, the recent policy debates on the environment and sustainable development are covered, together with their implications for transport. The second major change is the new government in 1997. The Conservatives had run out of ideas with the great transport debate of 1995 being translated into a non-committal policy statement, 'The Way Forward' (Department of Transport, 1996). Studies on road pricing in London and on motorway tolls had come to nothing, and the thinking from industry was that even if spending on new roads was raised by 50%, congestion would still get worse (CEBR, 1994).

However, no radical new ideas were present in the commitments made by the political parties in 1997. Statements were made on the railways, namely that they should be publicly owned and publicly accountable, but no mention was made about renationalization, only tougher regulation. Policy on roads was restricted to changes in the taxation system to encourage a modal shift away from the car, but road pricing did not figure prominently, except in the policies proposed by the Liberal Democrats. The major controversy seemed to be over the future plans for the London Underground, whether it should remain in public ownership or be privatized. The scene was set for new thinking, but it was unclear where that 'third way' might come from. In this Chapter, the development of the 'New Deal for Transport' is described, together with the evolution of Labour pragmatism.

Transport and the environment

Difficult choices have to be made. To accept the arguments for sustainability means a reduction in consumption and a much higher price for travel (Banister, 1993). It also means striking an appropriate balance between economic growth, the distribution of that growth and the environment. Environmental issues only form part of the political spectrum, and politicians have to balance economic growth with sustainable growth. Transport revenues form a significant part of total Exchequer revenues, and raising the cost of transport to the user to reflect the full social and environmental costs of travel may affect Exchequer revenues. Even if the price elasticity is low and people are prepared to pay more for travel, expenditure patterns in other sectors may be affected causing in turn a fall in demand, reduced growth and possible unemployment.

In 1990, the Department of the Environment published a White Paper, *This Common Inheritance* (Department of the Environment, 1990) which presented the first comprehensive review of every aspect of Britain's environmental policy. However, the options for transport seemed to be limited in their scope as they primarily related to reductions in energy consumption. The environment must be interpreted more broadly, as the environmental effects of transport influence all our lives in a variety of ways. Here are just two examples of the complexity of the interactions and the breadth of the transport and environment links.

In the past demand forecasts have been made for traffic and networks have been defined to meet that demand. It has now been realized that it may not be socially efficient, or desirable or possible to meet unrestricted demand and so restraint and management have become key concerns of transport planners. Alternative means of forecasting are being used such as scenario building, where alternative strategies are tested against possible future scenarios (Department of Energy, 1990). Alternatively, simulation studies have been developed to establish energy efficient forms of urban settlement patterns (Rickaby, 1987). In most cases these studies have taken one or two sectors (transport and land use) and examined environment in terms of one variable (energy). Yet no one has established the conditions under which a settlement pattern is energy efficient, let alone environmentally efficient. Transport factors have to be balanced against other energy costs such as the energy used in the construction and maintenance of the infrastructure (including buildings) together with the costs of space heating and ventilation. The energy factors have to be balanced against the economic costs of development, the availability and price of land, and labour costs. The qualitative factors which make up the environment add to this complexity. The research issue here is whether it is worthwhile to unravel these complexities and measure the interactions, or whether it is more sensible to take a modest perspective and examine policy questions individually.

The second example is that of health and stress as it relates to transport. In surveys carried out at the Marylebone Centre Trust (Tennyson, 1991) it seemed that transport was a major cause of stress both for users of the system and for those who live near to major transport routes. On the one hand, transport raises people's anger and aggression through such factors as the delays in traffic and actions of other drivers, whilst on the other hand transport creates fear and worry through the difficulties of crossing roads or travelling in very crowded conditions. Little research has been carried out on the effects of this on people's health, their absenteeism from work, their performance at work and their job satisfaction. In addition to the stress factors, public transport and certain spaces (tunnels and some roads) are perceived to be unsafe, and questions of security can help explain why some people are reluctant to travel after dark on their own (Atkins, 1990).

There is a wide range of 'solutions', yet little progress has been made to reduce the emissions from transport (apart from the compulsory use of catalytic converters on all new cars from 1993) or to reduce the consumption of fossil fuels. It is expected that emission levels and consumption of oil will continue to rise, with any benefits from increased efficiency being outweighed by the continual growth in traffic. If the UK was to achieve the target set by the Climate Change Convention (Department of the Environment, 1992) to return the emissions of each greenhouse gas to 1990 levels by 2000, action was required in all sectors. Yet in transport, the dilemma was clear as there seemed to be no obvious, politically acceptable means to reduce emissions and energy consumption.

This vacuum on environmental policy was filled by the Royal Commission on Environmental Pollution (RCEP, 1994) in their massive and influential report. After a comprehensive review of the environmental problems caused by transport, six options are presented:

- letting congestion finds its own level;
- predict and provide;
- greening the way we live;
- collective action;
- selling road space;
- relying on technology.

Very little guidance was given in the report as to how a strategy for sustainable transport should or could be put together, as it would require a combination of at least four of the options. The focus was clearly on the city, when much of the growth in traffic was taking place outside the city, on the motorways and the inter-urban road network, as well as in international travel. More fundamentally, little was said on the importance of persuasion and communication, as it is impossible to impose a radical solution, at least in a democratic society.

Public attitudes to the car must change if a sustainable transport policy is to become a reality. Technology would allow people to do what they are currently doing without using so many resources and creating less emissions. New fuels and low emission vehicles would help reduce the environmental costs of transport, and this option would always be acceptable as it requires little real change in travel. But even here the environmental arguments are not clear as to which fuel is the most appropriate long-term option. The Royal Commission seemed to favour natural gas, but higher quality petrol and diesel together with pre-heated catalytic converters might be as effective.

Pricing might make people more aware of the real costs of using the car, but current evidence suggests that many drivers are price resistant. Major increases in the costs of using the car are necessary if the intention is to get individuals to reassess the necessity of making a trip, of switching to the green modes (walk, cycle or public transport), or of travelling to a more local destination. Enormous revenues would be raised for the Exchequer and those might have a deflationary effect on the economy. If there were compensating actions to make the impact revenue neutral, then the impact would be reduced substantially. No one seems to have investigated these crucial macro-economic impacts. The expectation that a doubling of the real costs of fuel (1995–2005) will achieve the targets set for stabilization and reduction in CO_2 emissions is unproven, and further research is required on the effects (both direct and indirect) of a sustained real fuel price increase.

Ministers dismissed the Royal Commission's proposals as unrealistic, but its views may have influenced the introduction of the fuel duty escalator as the main means to achieve the greenhouse gas stabilization targets. It would also safeguard and increase Exchequer revenues from the transport sector. The escalator was originally set at 5% a year in real terms by the Conservatives, and continued by the new Labour government until 2000, when it was abandoned. The phasing out of leaded petrol seemed to be the only other response to the environmental debate.

The main contribution of the Royal Commission was to raise the quality of discussion on the environment and to set challenging targets for transport (Table 5.1). Such was the level of concern about the issue of transport and the environment, that the Royal Commission revisited the subject in 1997 to determine whether progress had been made (RCEP, 1997). Their conclusions were clear, namely that radical action was essential to make private transport less damaging and public transport more attractive, through creating an integrated transport decision-making process, through cleaner and more efficient vehicles, through incentives and market signals, and through more effective institutions. There had been little evidence of positive action over the intervening period.

Table 5.1 Summary of the transport and the environment objectives, together with the targets recommended by the Royal Commission

---

A: To ensure that an effective transport policy at all levels of government is integrated with land-use policy and gives priority to minimizing the need for transport and increasing the proportions of trips made by environmentally less damaging modes.

B: To achieve standards of air quality that will prevent damage to human health and the environment.

B1: To achieve full compliance by 2005 with World Health Organization (WHO) health-based air quality guidelines for transport-related pollution.

B2: To establish in appropriate areas by 2005 local air quality standards based on the critical levels required to protect sensitive ecosystems.

C: To improve the quality of life, particularly in towns and cities, by reducing the dominance of cars and lorries and providing alternative means of access.

C1: To reduce the proportion of urban journeys undertaken by car from 50% in the London area to 45% by 2000 and 35% by 2020, and from 65% in other urban areas to 60% by 2000 and 50% by 2020.

C2: To increase cycle use to 10% of all urban journeys by 2005, compared to 2.5% now, and seek further increases thereafter on the basis of targets to be set by the government.

C3: To reduce pedestrian deaths form 2.2 per 100,000 population to not more that 1.5 per 100,000 population by 2000, and cyclist deaths from 4.1 per 100 million kilometres cycled to not more than 2 per 100 million kilometres cycled by the same date.

D: To increase the proportions of personal travel and freight transport by environmentally less damaging modes and to make the best use of existing infrastructure.

D1: To increase the proportion of passenger-kilometres carried by public transport from 12% in 1993 to 20% by 2005 and 30% by 2020.

D2: To increase the proportion of tonne-kilometres carried by rail from 6.5% in 1993 to 10% by 2000 and 20% by 2010.

D3: To increase the proportion of tonne-kilometres carried by water from 25% in 1993 to 30% by 2000, and at least maintain that share thereafter.

E: To halt any loss of land to transport infrastructure in areas of conservation, cultural, scenic or amenity value unless the use of the land for that purpose has been shown to be the best practicable environmental option.

F: To reduce carbon dioxide emissions from transport.

F1: To reduce emissions of carbon dioxide from surface transport in 2020 to no more than 80% of the 1990 level.

F2: To limit emissions of carbon dioxide from surface transport in 2000 to the 1990 level.

F3: To increase the average fuel efficiency of new cars sold in the UK by 40% between 1990 and 2005, that of new light goods vehicles by 20%, and that of new heavy duty vehicles by 10%.

G: To reduce substantially the demands which transport infrastructure and the vehicle industry place on non-renewable materials.

G1: To increase the proportion by weight of scrapped vehicles which is recycled, or used for energy generation, from 77% at present to 85% by 2002 and 95% by 2015.

G2: To increase the proportion of vehicles tyres recycled, or used for energy generation, from less than a third at present to 90% by 2015.

G3: To double the proportion of recycled material used in road construction and reconstruction by 2005, and double it again by 2015

H: To reduce noise nuisance from transport.

H1: To reduce daytime exposure to road and rail noise to not more than 65 $dBL_{Aeq.16h}$ at the external walls of housing.

H2: To reduce night-time exposure to road and rail noise to not more than 59 $dBL_{Aeq.8h}$ at the external walls of housing.

Source: Royal Commission on Environmental Pollution, Press Release 'Transport and Sustainable Development' 26 October 1994

Transport and sustainable development

Two other major inputs to the debate were published in 1994, and taken together mark a major change in thinking on the environment and sustainable development. The UK Strategy for Sustainable Development provided a comprehensive review of all the elements that make up this complex concept and it presented the current situation (Department of the Environment, 1994). It was the UK government's response to the global stabilization targets set at the Rio Summit (1992), and it has now been superceded by the 1999 update (DETR, 1999a). The second document was PPG13 on Transport (Departments of the Environment and Transport, 1994), which brought together planning and transport through its three basic aims: to reduce growth in the length and number of motorized journeys; to encourage alternative means of travel which have less environmental impact; and to reduce the reliance on the private car. Again, this planning guidance has been recently updated (DETR, 2001).

The UK government over the last 10 years has been at the forefront of the international action over environmental pollution and the need to address the levels of consumption of non-renewable energy. Yet positive action seems to has been less apparent in the transport and planning spheres. The UK has more than achieved its targets for the stabilization of CO_2 emissions at 1990 levels, but transport has made no contribution as reductions have been made through switching from coal to natural gas for energy production, and through the natural process of de-industrialization. More difficult decisions have to be made, and this is where several new initiatives have taken place.

The DETR (1999b) states that

> sustainable development is about ensuring a better quality of life for everyone, now and for generations to come. It is concerned with achieving economic growth, in the form of higher living standards, while protecting and where possible enhancing the environment – not just for its own sake, but because a damaged environment will sooner or later hold back economic growth and lower the quality of life. Sustainable development is equally concerned with making sure that these economic and environmental benefits are available to everyone, not just to a privileged few.

The challenge for transport planners is to bring such a concept to reality. This requires a fundamental rethinking of the role of planners as guardians of the public interest, so as to become much more actively involved in the debates over the future of town and country. Transport planning and planning more generally must move from being a control function to a more proactive initiator of change so that the vision of sustainable development can be achieved. Apart from a fundamentally different approach to development, the necessary skills, powers, resources and organizational changes also need to be made. This is where the difference between the rhetoric and the reality becomes apparent, and good intentions never really get effectively implemented.

Although it has been argued that transport planning is central to sustainable development, the concept itself is far wider. The UK Sustainable Development Commission has focused the debate on six main elements (Owen, 2001):

* access including transport, mobility and the planning/land-use dimension;
* improving eco-efficiency with a particular focus on production processes;
* health, including decoupling it from sickness treatment and the role of lifestyle and consumption choices;
* food production and consumption, with a focus on agriculture and rural land use, including environmental management of the rural landscape;
* participation, leadership, citizenship and government;
* joined up government and the evidence based approach.

Planning impinges on several of these issues, but we will focus on the first, together with some reference to the other factors where necessary, principally in 5 and 6.

Underlying much of the debate on sustainable development is the importance of accessibility or the ability to reach a range of services and facilities easily. This simple concept is at the heart of much planning and transport thinking as its intention is to keep travel distances as short as possible. One of the fundamental axioms of transport planning is that people do not like travelling, and they only travel because the benefits received at the destination (for example, work, education) more than outweigh the costs (i.e. time and money) of getting there. An important corollary of this axiom is that if journey distances are short, then people are more likely to use 'green modes' of transport (for example, walk and cycle) or public transport. As journey lengths increase so does the likelihood that the car will be used. The second corollary here is that the solution is not really a transport one, but a land-use and development one. If decisions are made to locate new housing, shopping, employment, recreational and other facilities in close proximity to each other, then journey lengths should be reduced as people have the opportunity to do things locally. This solution will maximize accessibility, and through integrating both transport and planning will substantially improve the potential for sustainable development.

However, it is also recognized that location strategy alone will not necessarily mean that people will use those local facilities. For example, they may choose to travel further to the supermarket, rather than use the local shop. The debates over participation and citizenship must be undertaken and supported, so that individuals are aware of the consequences of their choices for sustainable development. We are all part of the problem and so we must also be part of the solution. The accessibility strategy outlined here meets the environmental concerns of sustainable development and the equity concerns. For those people with no car available, local facilities mean that their levels of accessibility are high. Although increased transport services can be provided to give choice to people, the most important determinants are where the facilities and services are themselves located. This is where land-use planners and transport planners must work closely together.

The evidence is also clear, namely that the car has increased its dominance at the expense of all other modes of transport. Even the healthy green modes have suffered. The goal of sustainable transport is an illusion as all the transport indicators are in the wrong direction. Transport's contribution to sustainable development over the recent past has been negative. Even if we take other indicators, such as energy use and levels of some major emissions, the same conclusion is reached. Transport now consumes 61% of all petroleum products in the United Kingdom and is a significant contributor to carbon dioxide (25%) and other emissions (see Table 3.7). Technology has a major role to play here through add on technology, principally the catalytic converter and particulate traps, and through more efficient engine design. However, there is no known technology that will reduce CO_2 emissions which is transport's main contribution to global warming. If carbon based fuels continue to be used, then CO_2 will be produced in ever increasing quantities from transport, unless technological innovation radically improves engine efficiencies and accelerated replacement of old vehicles takes place.

One of the main reasons for increasing car dependence and journey lengths has been the closure of many services and facilities, together with consolidation and specialization within outlets. Often there are substantial cost savings to the provider of the service (for example, the health service or the local education authority) as there are substantial economies of scale. But some of these costs are passed on to the users of the service who have to travel further to get to the facility. A social audit on proposals to close a facility would establish the overall welfare gain or loss from the closure. That audit would form part of the decision as to whether that facility should stay open or not, and a subsidy could be paid to the provider of the service to keep it open (Banister, 1997a).

Similarly, the social audit of new development would assess the benefits to the local economy and users of the new facility, but also the additional costs in terms of travel distances and impact on the town centre. It extends the notion of the sequential test[1] being used for new shops and residential development, as local authorities could place restrictions on the planning permission, so that the developer had to ensure that a certain proportion of all trips were made by modes other than the car. The means by which this would be achieved would be left to the developer, but failure to reach the target levels would result in fines. Similarly, a development levy could be charged for development on greenfield sites so that the proceeds could be used to help in remediation on brownfield[2] sites in urban areas. At present it is far cheaper to build new houses (VAT exempt) than to refurbish or convert the existing stock. The VAT rate for conversions was recently (November 2000) reduced from 17.5% to 5% (DETR, 2000b), but this revenue will not be reinvested in urban regeneration projects as suggested by the Urban Task Force (1999, p. 254).

In terms of planning, the principles of sustainable development are clear, namely that higher densities are required and that most new development should be placed in settlements of a sufficient size to support a full range of services and facilities, possibly with mixed land uses (Table 5.2). In transport terms, such development will reduce the growth in car based travel, energy consumption and emissions levels. In addition to the physical aspects of planning, it is important to maintain strong communities through increasing local economic diversity, through increasing self-reliance, through reducing the use of energy, through the protection and enhancement of biological diversity, through the stewardship of natural resources, and through social justice (for example, by providing for the housing and living needs of all). Such a series of claims may be idealistic, but the attractiveness of any location should be promoted through investment in high quality public services. This involves the public sector working in partnership with the private sector and voluntary organizations.

The new millennium has marked a watershed in thinking from central government with the production of two White Papers, one on towns and cities and the other on the countryside (DETR, 2000c, 2000d). They are really a response to the challenging vision of an Urban Renaissance described by the Urban Task Force (1999), and summarized in Table 5.3. The urban areas[3] are seen to be central to sustainable development as some 80% of people live here and most new development must be accommodated within these areas.

Table 5.2 The principles of sustainable development in planning and transport

Land Use	Settlement size – over 50,000 Density – over 40 dwellings per hectare Mixed land uses Proximity to public transport interchanges within 500 metres of a bus stop
Quality	Open space and quality in urban design Safe and secure environment Peace and quiet Social and recreational opportunities Full range of local services and facilities

The government has clear targets for 60% of all new housing to be located within existing urban areas, mainly on brownfield land, and to be developed at higher densities. This imperative is highlighted by the requirement for 3.8 million new dwellings in England by 2021, mainly to accommodate the huge growth in single-person households. Conversion and new construction in the cities will help to reduce the migration of people out of the cities, but it is crucial that the new small units are affordable.

There is a 'time bomb' effect here, as many people in the public sector cannot afford to live in the South-East or in London, yet these people are essential to the city economy and the efficient operation of cities. If the problem of pay and affordable housing is not addressed, then many health, education, social and public services, and even government functions may break down. And business and industry also depend on these employees to service their activities. In the past, the solution has been that the public housing sector has accommodated these workers, but increasingly they have become excluded from the housing market and have moved out to the suburbs so that cheaper housing is matched by longer (and more expensive) commuting. The housing market has failed, so positive investment is essential to address the lack of affordable, small units not only in urban but also rural areas. Yet, both of the White Papers have seriously underestimated the scale and severity of this problem (DETR, 2000c and d).

The housing problem also reflects regional inequalities and the juxtaposition of wealth and poverty within urban and rural areas. The Regional Development Agencies (RDAs) are primarily concerned with inward investment and the creation of jobs, as this is seen as the means to improve competitiveness. Such strategies do not address issues relating to skills, learning and education, or to those relating to social inclusion, or enhancement of the environment. It is solely the economic imperative. The RDAs need to broaden their remit to encompass local needs and to combine all these issues equally. In turn, this means that planning has to be reformed so that it can encourage or discourage development according to its overall resource intensity, including its impacts on the environment and on local communities. Investment and taxation should be related to a wider concept of employment to cover range of jobs, the quality of jobs and new types of jobs (for example, job sharing). This in turn may require a wider notion of the city within its region, perhaps defined by the journey to work areas so that the tax base reflects this wider geography.

However, despite these good intentions, policy thinking on transport and planning has not changed fundamentally to embrace the new framework for sustainable development. The *Ten Year Plan* for transport (DETR, 2000b) has turned the clock back to renewing the priorities for the large-scale projects and road building. The notions of accessibility and reducing travel distances have been forgotten as the dominance of the car driver is re-imposed, with mobility and choice being re-established as the key concepts, and the main objectives is to reduce congestion. But even here, the *Ten Year Plan* does not deliver, as congestion is seen as relating primarily to travel speed, not to the instability and uncertainty in travel time, which is the main problem facing all travellers (CPRE, 2001). Perhaps it is optimistic to expect radical change over such a short period of time, but unless new thinking and action is forthcoming, the quality of the environment will continue to deteriorate. Perhaps it is through direct action that change will take place, as people become more interested in the issues and frustrated with the lack of effective action.

Recent evidence from transport disasters, fuel price protests and adverse weather conditions has tested the resilience of the transport system to its limits. In the past, the argument has always been about increasing speed to save time, and most benefits from transport investments have been travel time savings. Yet this rationale has also led to increasing journey lengths and more travel. Perhaps the new sustainable development agenda would argue for a slowing down of travel speeds to encourage shorter journeys and the use of local facilities. This would include the closure of some roads at certain times so that they can be returned to the people. Such changes have happened in urban areas with the 'Reclaim the Streets' campaign where, for a short period of time, people have taken control of the street space for themselves rather than traffic. But these measures are equally applicable in the countryside, as roads could be allocated to school children for cycling or to ramblers for walking, at least for part of the day or for the summer. At other times, these same roads could be opened to more general traffic, but with much lower speed limits.

Concepts such as reasonable travel time must become part of the policy agenda. It increasingly seems that the transport network is only as good as its weakest link. At present high-speed networks that encourage longer distance travel are more vulnerable to failure, particularly if the organizational structure is fragmented (as with the railways) and many services are outsourced (as in the freight sector). These weaknesses may be compounded by too great an emphasis on commercial criteria (for example, profit) as the main measure of efficiency. Broader social criteria also need to be considered as to who gains and loses from particular decisions, and the consequences for environmental quality.

Sustainable development must be seen as a distinctive and necessary policy agenda, rather than an add on to the economic agenda. At present, the two White Papers (DETR, 2000*c* and *d*) are embedded in the Treasury driven agenda of economic growth narrowly defined, and this will not lead to sustainable development. The debate must be matched across all sectors with the availability of affordable housing, local facilities, and high quality local

Table 5.3 The Urban Renaissance

Theme	Proposals
1 The sustainable city Establishes the importance of developing a higher quality urban product by creating compact urban developments, based upon a commitment to excellence in urban design and the creation of integrated urban transport systems that prioritize the needs of pedestrians, cyclists and public transport passengers.	• Create a national urban design framework, disseminating key design principles through land-use planning and public funding guidance. • Undertake area demonstration projects which illustrate the benefits of a design led approach to the urban regeneration process. • Make public funding and planning permissions for area regeneration schemes conditional upon the production of an integrated spatial masterplan. • Commit a minimum of 65% of transport public expenditure to programmes and projects which prioritize walking, cycling and public transport, over the next ten years. • Place local transport plans on a statutory footing. They should include explicit targets for reducing car journeys and increasing year on year the proportion of trips made on foot, bicycle and by public transport. • Introduce home zones, in partnership with local communities, which give residential areas special legal status in controlling traffic movement through the neighbourhood.
2 Making towns and cities work Improving the management of the urban environment, targeting resources on the regeneration of areas of economic and social decline, and investing in skills and innovative capacity.	• Give local authorities a strategic role in managing the whole urban environment, with powers to ensure that other property owners maintain their land and premises to an acceptable standard. • Create designated Urban Priority Areas, where special regeneration measures will apply, including a streamlined planning process, accelerated compulsory purchase powers and fiscal incentives. • Develop a network of Regional Resource Centres for Urban Development, promoting regional innovation and good practice, co-ordinating urban development training and encouraging community involvement in the regeneration process.
3 Making the most of our urban assets Developing on brownfield land and recycling existing buildings must become more attractive than building on greenfield land. The priority is to make the planning system operate more strategically and flexibly in securing urban renaissance objectives in partnership with local people.	• Make statutory development plans more strategic and flexible in scope, and devolve detailed planning policies for neighbourhood regeneration into targeted area plans. • Produce dedicated Planning Policy Guidance to support the drive for an urban renaissance. • Adopt a sequential approach to the release of land and buildings for housing, so that previously developed land and buildings get used first. • Require local authorities to remove allocations of greenfield land for housing from development plans where the allocations are no longer consistent with planning policy objectives. • Establish a national framework for dealing with the risks that arise throughout the assessment, treatment and after-care of contaminated sites. • Require every local authority to maintain an empty property strategy that sets clear targets for reducing levels of vacant stock. • Establish a Renaissance Fund whereby community groups and voluntary organizations can access the resources needed to tackle derelict buildings and other eyesores spoiling their urban neighbourhood.
4 Making the investment Sufficient public investment and fiscal measures must be used to lever in greater amounts of private investment into urban regeneration projects.	• Establish national public-private investment funds and regional investment companies, to attract additional funding for area regeneration projects. • Introduce a new financial instrument for attracting institutional investment into the residential private rented sector. • Introduce a package of tax measures, providing incentives for developers, investors, small landlords, owner-occupiers and tenants to contribute to the regeneration of urban land and buildings. • Include the objective of an urban renaissance in the government's spending review which will determine public expenditure priorities for the early years of the new millennium. • Review local government spending formula, which determines the allocation of central government resources, so that it reflects the financial needs of urban authorities managing and maintaining their areas.
5 Sustaining the renaissance New apparatus will be required to ensure that the goal of an urban renaissance remains a political priority over the 25 year period of the household projections.	• Publish an ambitious Urban White Paper, which addresses economic, social and environmental policy requirements, tying in all relevant government departments and institutions. • Establish an Urban Policy Board which combines national, regional and local leadership in driving the renaissance at all levels of government. • Introduce an annual 'State of the Towns and Cities' report to assess progress against key indicators. • Create a Parliamentary Scrutiny Committee to ensure government accountability for the delivery of urban policy objectives.

Source: Based on the Urban Task Force (1999), pp. 11–12

transport networks. This is the essence of the accessibility argument that reduces dependence on resources and promotes a high quality of life.

One means to overcome the lack of integration between the various policy concerns of government has been the development of a series of national indicators of sustainable development, together with 15 key indicators to give a broad view on whether a better quality of life is being achieved (DETR, 1999*a*). The UK government has taken this bold step to highlight quality of life through monitoring progress on all indicators, and to take action where the trend is unacceptable (Table 5.4). It is early days, but most of the changes in the 15 headline indicators are not measurable yet, as we have indicated, the trends are still continuing, mainly in the 'wrong' direction. The difficulty is that we know what is happening, but we do not really want to change the direction. The indicators will not help change direction, but only reinforce the message that we are not moving towards sustainable development. Note also that the use of indicators is not as demanding as the use of targets such as those set by the Royal Commission on Environmental Pollution and summarized in Table 5.1.

Ironically, it is only in the economic indicators (Table 5.4, 1a, 1b, 1c) that steady positive progress is being made. Yet these indicators provide no information about sustainable economic development, as they are standard measures of economic growth. Some progress is also identifiable in the social indicators (Table 5.4, 2a–2e), but it is in the environmental indicators (Table 5.4, 3a–3f) that least progress is being made, particularly in the longer term where tougher targets are needed to address the global issues of sustainable development (RCEP, 1994 and 1997).

For example, the recent Intergovernmental Panel on Climate Change report (Watson *et al.*, 2001) has proposed a 60% reduction in global greenhouse gas emissions as its best estimate of the scale of change required by 2050 to limit the adverse impacts on the global climate. This means that the developed countries should aim at a target of about a 90% reduction! But the UK government has not moved further than its 20% target reduction for CO_2 emissions, which in itself is one of the most challenging targets set by any country. Its most recent set of proposals for more roads, no traffic reduction targets, the relaxation of the fuel duty escalator have all suggested that even this 20% target is going to be impossible to achieve, at least in the transport sector. There are no clear national targets in planning guidance, with the only possibility that the local authorities themselves will impose local targets, but they are naturally reluctant to do so individually. As the CAG Consultants (2001) report concludes, 'taken as a group, the indicators provide no evidence that the UK has moved significantly closer to reconciling and integrating quality of life with living within environmental carrying capacities'.

Underlying all of the rhetoric is the suspicion that sustainable development is not really the focus of policy, but a 'side show'. The economy itself is the most important concern of all regions and governments, and it takes precedence over all other interests. The increasing reliance on the market, albeit modified by social concerns, is paramount. In second place, a long way behind, are the real concerns over social welfare and an inclusive society, with environmental issues lagging behind in a very poor third place. Unless this balance is redressed, sustainable development is an impossibility, and the indicators used will merely reflect the pre-eminence of economic and some social measures over and above the others. This conclusion is illustrated by reference to the five headline indicators of relevance to planning and transport, where comments have been inserted in bold in Table 5.4 (indicators 2d, 3a, 3b, 3c, and 3f).

The use of indicators can help in identifying where quality of life is improving or getting worse, but they should only be used in that capacity. The risk is that too much effort is expended in monitoring trends that we already know about, rather than actually tackling the problems at source. This is where the vision of the future is required, and policy action taken. The indicators do not say anything about why a change has taken place, as they cannot be linked to causal factors. These is a real danger that 'indicator watching' is merely observing the symptoms rather than the underlying causes of changes in the quality of life. It would be far more effective to develop real targets, combined with a vision of a future, that covers sustainable development at the national, regional and local levels, and to achieve them over a given time period.

Quality of life is a much more fundamental concept than that suggested by these indicators. At the individual level, it relates to satisfaction, self-fulfilment, security, health, independence, confidence, and the challenges of life. At the community level, it relates to the quality of the built environment, the neighbourhood, social diversity, and the local natural environment. At the national level, it relates to economic prosperity, political stability, environmental quality, and a fair society. It is impossible to combine all these elements into a series of simple quality of life indicators.

A new element is emerging in the discussion, namely the adverse health implications of transport. This manifests itself in many ways through physical and mental health, through reductions in the levels of physical activity of the population (particularly the young), and through the steady reduction in the use of bicycles and walking as modes of transport. Again, this affects the quality of life and its importance is often underestimated. The increased dependence on the car has resulted in policies that rely mainly on curative solutions (for example, technical and engineering alternatives), rather than preventative solutions, to make the medical analogy. The main concerns relate to health and pollution, road deaths and injuries, noise and the

Table 5.4 The headline indicators in the UK sustainable development strategy

Themes, issues and objectives	Headline Indicators
1 Maintaining high and stable levels of economic growth and employment	
1a The economy must continue to grow	Total output of the economy (GDP and GDP per head)
1b Investment (in modern plant and machinery as well as research and development) is vital to our future prosperity	Total and social investment as a percentage of GDP
1c Maintain high and stable levels of employment so that everyone can share greater job opportunities	Proportion of people of working age who are in work
2 Social progress which recognizes the needs of everyone	
2a Tackling poverty and social exclusion	Indicators of success in tackling poverty and social exclusion (children in low-income households, adults without qualifications and in workless households, elderly in fuel poverty)
2b Equip people with the skills to fulfil their potential	Qualifications at age 19
2c Improve health of the population overall	Expected years of healthy life
2d Reduce the proportion of unfit (housing) stock	Homes judged unfit to live in – **some limited progress here with the % falling from 8.8% (1986) to 7.2% (1996)**
2e Reduce both crime and people's fear of crime	Level of crime
3 Effective protection of the environment	
3a Continue to reduce our emissions of greenhouse gases now, and plan for greater reductions in longer term	Emissions of greenhouse gases – **Reductions of 8% in CO_2 emissions (1990–1998), mainly due to closure of coal power stations and replacement by nuclear and gas power stations. Levels expected to rise again, principally from transport – 20% reduction target for 2010 very difficult.**
3b Reduce air pollution and ensure air quality continues to improve through the longer term	Days when air pollution is moderate or higher – **Reductions in urban areas from 59 days in 1993 to 23 days in 1998 due to fall in particles and sulphur dioxide. Corresponding levels in rural areas fluctuate between 21 and 50 days between 1987 and 1998, due to particles and ozone.**
3c Improve choice in transport; improve access to education, jobs, leisure and services; and reduce need to travel	Road traffic – **Car distance travelled has increased by 65% per person (1980–2000) and this is likely to continue, but at a lower level – perhaps a 35% increase to 2020. National traffic reduction targets have not been set and the aim is only to reduce levels of expected growth.**
3d Improving river quality	Rivers of good or fair quality
3e Reverse the long-term decline in populations of farmland and woodland birds	Populations of wild birds
3f Re-using previously developed land, in order to protect the countryside and encourage urban regeneration	New homes built on previously developed land – **Some progress has been made towards the national target of 60% of new dwellings on land already used (brownfield development) – 1990 = 50%, 1997 = 55%, 2008 = 60%**
4 Prudent use of natural resources	
4a Move away from disposal of waste towards waste minimization, reuse, recycling and recovery	Waste quantities and management

Source: DETR (1999a) www.environment.detr.gov.uk/sustainable/quality99/chap3

fragmentation of communities (Fletcher and McMichael, 1997). In many cases, the evidence is still open to interpretation and the crucial issues of causality (the actual link between transport and health), the key trigger mechanisms (for example, the asthma attack), and the effects of pollutants in combination (the 'cocktail' effect) are not well known. But there is sufficient evidence for the precautionary principle to be followed to limit the potential damaging effects on health. Increasingly, these factors are now considered important in the evaluation of transport projects (see Chapter 3) and in the quantification of the externalities associated with transport (Maddison *et al.*, 1996).

Within the broader debate on sustainable development, there is a hope that health might play an instrumental role in changing public attitudes. The White Paper on Transport Policy (DETR, 1998a) makes a strong case for a 'healthy nation' through reductions in car use, through increases in walking and cycling, and through the use of the best available technology. Coronary heart disease is the largest killer of adults in the UK, and it is estimated that up to 24,000 vulnerable people die prematurely each year, and a similar number are admitted to hospital, because of exposure to air pollution, much of which is due to road traffic (DETR, 1998a). Car drivers themselves are not exempt as they face pollution levels inside a car two or three times higher than those experienced by pedestrians! All those travelling are at risk (as well as those living near to busy roads), and policy should be designed to reduce those risks to acceptable levels for all. At present the balance is too much in favour of car drivers (BMA, 1997). The opportunity is present for change through the requirement for real collaboration between those seeking to promote preventative health and those responsible for transport planning. It is not just the direct effects that are important, but the indirect effects, in terms of feelings of isolation and insecurity, of anxiety and fear of going out, that need to be addressed if quality of life (and health) is to be improved.

The third way

The Labour party was returned to power in 1997 after some 18 years watching radical change in transport policy. They were full of new ideas and enthusiasm, with a strong personality in the restructured 'mega' Department of the Environment, Transport and the Regions (DETR – Table 5.5). All the signs were positive, yet even this combination has found the going tough, particularly in the first 3 years when the Treasury pursued a continuation of the policy of tight fiscal control (Table 5.7).

The same three issues have been at the forefront of the new thinking on transport policy, namely congestion (roads and air), privatization (rail), and investment in new and replacement infrastructure. However, the packaging

Table 5.5 The Labour government 1997–2001

1997–2001	**May: Labour elected majority 179** **Tony Blair – Prime Minister** Department of the Environment and the Department of Transport combined into a new Department of the Environment, Transport and the Regions – June 1997 Agreement to cut six main greenhouse gases agreed in Kyoto, Japan – December 1997 White Paper on Mayor and Assembly for London – June 1998 White Paper on Integrated Transport Policy – July 1998 EU ministers agree to cut CO_2 emissions for new cars by 25% in 2008 – October 1998 Shadow Strategic Railway Authority launched – April 1999 Transport Bill published – December 1999 Jubilee Line Extension opened – December 1999 Croydon Tramlink opened – May 2000 Mayor for London elected – May 2000 Transport 2010: The Ten Year Plan published – July 2000 Hatfield rail crash – October 2000
2001	**June: Labour re-elected majority 167** Department of Transport, Local Government and the Regions – June 2001

Note
This table is a continuation from the Conservative governments in Table 4.1

was very different as integration in transport became a prime concern and central to the transport policy statement of 1998 (DETR, 1998*a*) where ten key points were highlighted (Table 5.6). Transport was at last being seen as part of a much wider set of public service activities where accessibility was regarded as central. This meant looking at ways in which the best use could be made of all forms of transport through multi-modal journeys and convenient interchange,

Table 5.6 The Transport White Paper 1998 – the 10 key points

1   Charges by 2005 for driving into town centres and for workplace parking – money to be reinvested in public transport;

2   Local transport plans (LTP) to be produced for the following 5 years;

3   Strategic Rail Authority (SRA) with more powers to fine Railtrack and the train operating companies for failing to invest or meet targets. SRA to negotiate with train operators on new franchises and to have responsibility for freight strategy;

4   Commission for Integrated Transport to advise and monitor progress;

5   Voluntary quality partnerships between local authorities and bus companies to deliver better services;

6   National public transport information system by 2000;

7   Safer routes for cyclists and walkers;

8   National concessionary fares scheme – half price tickets for pensioners;

9   More resources allocated to encourage freight to switch from road to rail – but road hauliers to be allowed to operate 41 tonne vehicles on 6 axles;

10 Privatization of the air traffic control system.

through ticketing, high quality real time travel information, and attractive public transport. It also extended the concept of accessibility to include education, health and other services, so that transport becomes a key element in wealth creation and an inclusive society. In the revised PPG13 on transport (DETR, 2001), all new development proposals, that had a significant transport impact, will have to submit a transport accessibility assessment to determine whether safe and easy access can be guaranteed for those with no access to a car. It is not just accessibility that is important, but the total numbers of trips attracted to and generated by new development. Integrated transport policy should not just consider the accessibility requirements, but all the transport implications of different development strategies.

However, the reality has been different with central government expenditure being limited to £6000 million on transport over its first 3 years (1997–1999 – Table 5.7), which is probably about 20% of what is required. Resentment amongst car drivers was increasing as the fuel duty escalator continued and increased, and as congestion got worse the view was that the motorists were not getting a good deal (Table 5.8). The £6000 million spent on transport was only a small proportion of the £33 billion raised in taxes from road users (1998). This public resentment has been reflected in the tension within the government between the politically aware Prime Minister who needs the support of the motorist and the more idealist Secretary of State who was committed to the promotion of public transport. Progress has been made with a relaxation of Treasury constraints on public expenditure and the

Table 5.7 Central and local government expenditure on transport, Great Britain (£ million)

	1996/97	1997/98	1998/99	1999/2000
Total	6251	5930	5621	5948
National roads	2018	1840	1703	1755
Local transport	4233	4090	3918	4193
Capital	1284	1282	1099	1259
Current	2949	2808	2819	2934
Concessionary fares	480	460	490	540
Revenue support for public transport	551	529	538	538
National support for rail[1]	1874	1438	1207	1111
All expenditure	8715	8693	8048	8817
Real expenditure (1999/2000 prices)	9665	9362	8306	8817

Notes
1 Includes both national rail and London Underground
Total = national roads and local transport
Local transport = capital and current
Current = revenue support, concessionary fares and others
All expenditure = total, rail and other
This table is a continuation from the Conservative governments in Table 4.2
Sources: DETR (2000a) and www.wolfbane.com/rpi.htm

possibility of hypothecation of some transport charges. This important concession has now paved the way for revenues raised from road pricing and workplace parking levies to be used for investment in public transport and roads schemes. This change was crucial to the implementation of the new policy agenda, as public acceptance is significantly increased if those revenues are targeted at transport. It has yet to be seen whether they will be seen wholly as a net addition to the transport budgets of local authorities, or whether there will be a readjustment in other sources of funding to moderate that net addition.

The more basic questions on transport policy have not entered the debate. Fundamental here, particularly in light of the Royal Commission's Report (RCEP, 1994 and 1997), was the means by which the car should be accommodated within the city, and the role that it should play. There is a general recognition of the problem of matching supply and demand, and the options are well known (Banister, 1997a). Most car users support restrictions and do not want new roads, particularly near to where they live. But they still wish to own the cars they use, and to use them where and whenever they want. There is also an increasing resistance about paying more for that privilege, particularly among low-income households (including many rural poor) and small businesses (including owner drivers of lorries).

The House of Commons Select Committee on Environment, Transport and Regional Affairs (1999) has come up with a range of clear proposals and a strong request for more cash:

◆   Tougher planning rules prohibiting edge and out of town developments;
◆   New roads should only be built where large economic benefits can be shown to follow;
◆   New bus regulator, whole route approach to bus priority, and comprehensive information for all forms of public transport;
◆   More paths and cycleways, including better personal security and CCTV at stations;
◆   Target of 50% cuts in road deaths and serious injuries by 2010;
◆   More investment by the Strategic Rail Authority and train operating companies with penalties if targets are not met. Access charges to be split into operating and investment charge elements, with the former going to Railtrack and the latter to the SRA.

Yet with all of these proposals and the key points raised in the Transport White Paper, there is still a strong sense that the optimism of new Labour has been dented. They are struggling to find a third way when it comes to transport. It is a policy minefield that they would like to ignore, but they cannot. There was no fundamental change, only the imposition of more controls on the market approach introduced by their predecessors. But then

Table 5.8 Increasing levels of congestion in Great Britain

	Cars per kilometre of road	Vehicle-kilometres per kilometre of road (thousands)
1967	57.7	1136
1977	84.0	1568
1987	108.7	2187
1997	129.0	2748
2010	169.1	3405
2025	200.0	3795

Notes
Roads measured in kilometres of classified roads – motorways, trunk roads, Class 1 or principal roads, Class 2 or B roads, and Class 3 or C roads
Forecasts based on National Road Traffic Forecasts (DETR, 1997c)

again, there seem to be strong limitations on how effective new controls would be, in particular as they relate to the privatized rail system.

The public sector alternative used in France (Chapter 7) seems to offer an attractive option, but there is the continuing problem of catching up with the backlog of investment. The Labour government's investment plans have been to continue those put in place by the previous Conservative government. But now, there is a backlog of some 30 years under-investment to catch up on. The expectation that the private sector would 'fill' the gap is unrealistic (Chapter 3), as there are many good reasons for not investing in transport. Even the possibilities offered by the Private Finance Initiative (PFI) and Public Private Partnerships (PPP) have not succeeded in bridging the funding gap, as even here the risks are seen to be too high, and the returns are not guaranteed or are too low (Chapter 10). Successive governments have fallen into the same trap, and in the meantime levels of congestion have got worse, as the growing economy has generated more travel (Table 5.8). In the short and medium terms, under-investment in the transport infrastructure is an attractive option as services continue to operate with little apparent deterioration in quality. But in the longer term, the costs of repair, replacement and new investment continue to mount up.

This impending disaster was the background against which the 10 year plan was generated. For the first time a vision of the transport situation in 2010 was given, together with a massive £180 billion programme of investment, split equally between roads, railways and local transport. The strategic plan for transport is a breakthrough, as it takes planning out of the annual Treasury cycle which is determined by short-term macro-economic conditions, to placing investment on a longer term horizon (Table 5.9). Such thinking is essential to the rebuilding of the UK infrastructure.

However, when the headline numbers are broken down, the public sector contribution shrinks to £132 billion, with the balance being sought from the private sector. Even then, about half of the total budget will be allocated to

Table 5.9 The background analysis and responses in the transport 2010 plan

1   Traffic growth to increase by 22% and congestion[1] to increase by 15% (and 28% on the trunk road network) between 2000 and 2010. Plan to reduce this by 5% through bypasses, elimination of targeted bottlenecks, local schemes and 40 major road schemes.

2   Overcrowding levels of commuting by rail is 40% and congestion in London is 3.5 times the England average. Plan to reduce this through new investment in Crossrail and tram systems in London and through new capacity on the Underground.

3   Inadequate public transport in England, which has not been resolved by deregulation. Plan to increase use of buses by 10% and to build 25 new tram systems.

4   Rural transport is not sufficient to meet the needs of those without access to a car. Plan to extend rural bus subsidy grants, fuel duty rebates to be extended to community transport, and to improve quality of information.

5   Substantial maintenance backlog of the roads. Plan to eliminate this through a £30bn programme.

6   Rail demand will increase by 34% between 2000–2010, but capacity restraints will limit this to 23%. Plan to improve services and to target a 50% increase in patronage.

7   Freight distribution needs to be made more efficient as it accounts for about 10% of GDP. Plan to increase rail freight by 80%.

8   Safety record on roads is good, but less so on the railways. Plan to reduce by 33% the numbers killed or seriously injured on the strategic road network by 2010.

9   Climate change targets will be increased from the 4.0 million tonnes of carbon (MtC) from the voluntary agreement with the industry to 5.6 MtC with the plan. No comment made as to how.

10 Air pollution and noise to be reduced by 50% by 2010, through improvements in vehicle technology and increased fuel economy and quality.

11 Social exclusion to be reduced. No comment made, but implied in the measures to be implemented.

12 Passenger numbers through airports to increase by 50% to 2010.

Note
1   Congestion was defined as the delay experienced for each kilometre travelled compared with when traffic is light

subsidy for the railways and routine spending on road maintenance, leaving £65 billion for new investment. The expectation of substantial private funding is mainly for the national railways and London Underground, but the increased uncertainty (post the Hatfield Rail disaster) and the doubts over the future of the underground system in London means that the private sector will be even more reluctant to invest. The debate needs to be held as to whether privatization is the long-term future for the railways. Ownership of the infrastructure, together with the primary responsibility for investment in that infrastructure, increasingly seems to lie in the public sector. The role of the private sector should be to run high quality services and to act as the public face of the transport system as it interfaces with the user. This debate still needs to be held, despite the recent growth in the use of the national railways (Figure 5.1). It should be remembered that 70% of passenger expenditure on rail comes from the top 40% of earners, and that on average only 12 rail trips

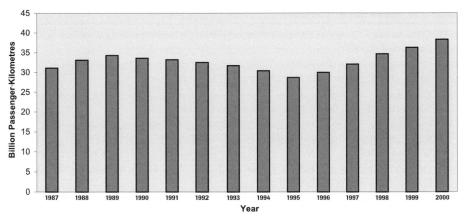

Figure 5.1 Rail patronage 1987–2000

are made by each person each year. This means that the distribution of use is heavily skewed to certain high users who commute, while others make no use of the railways.

The 10 year strategy lacks the coherence that the 1998 White Paper had, and it fails to build the integrated transport system. The strategy seems to divide the total financial resources into equal slices for each part of the system (the Salami approach). Essentially, it goes back to the modal basis for transport policy, which treats each form of transport as separate and in competition with each other. It does not attempt to integrate transport in trying to make the best use of the system as a whole and to break down the traditions of a mode driven policy.

Conclusions

Transport policy seems to lack any clear direction and there is always an underlying feeling that despite good intentions, a new phase of road construction is imminent (Table 5.10). The impasse is essentially a political one as policies which might address the congestion and environmental issues are available, but they do not seem to be politically acceptable. As Vigar (2001) suggests, there is no coherence in the discourse and no consistent storyline. In the past, the approach has been through consensus building with strong financial direction from the centre, moderated by local discretion. Some progress has been made through the concept of integration within transport and between transport and other sectors (the 1998 White Paper), and through the recognition that the user should pay the full costs of their travel, but at the same time being aware of particular social needs and the requirement to maintain and improve levels of accessibility. Demand

Table 5.10 Roads policy 1989–2001

1989	Conservative – £23 billion Roads for Prosperity: Paul Channon 'the biggest roads programme since the Romans'.
1992	Conservatives – Fuel duty escalator – commitment to raise duty by 5% per annum in real terms – Public protest – M3 Motorway through Twyford Down.
1993	Conservatives – PPG6 guidance to reduce out of town retail developments – sequential test – New road through Oxleas Wood in south East London dropped.
1996	Conservatives – predict and provide ended. No longer to build roads to meet demand. Public protest – A34 Newbury Bypass. Roads review abandoned 360 out of 500 roads in the Channon 1989 Plan.
1997	Labour – Raises fuel duty escalator to 6% per annum. New bypass for Salisbury dropped.
1998	Labour – First Statement on Transport Policy for 20 years – introduces urban road pricing, the Strategic Rail Authority (Table 5.6). No statement on roads but limits on investment.
2000	Labour – 10 Year Plan – £60 billion over ten years on new roads and maintenance.
2000	Labour – Fuel duty escalator dropped (March) and fuel price protests (September).
2001	Labour – PPG13 extends the sequential test to all development with clear imperative to locate development in town centres or at transport interchanges, particularly where there will be high levels of traffic demand (similar to Dutch ABC policy – Chapter 7).

management has been accepted within urban areas, but not on the interurban road network or at attractive leisure locations. The optimism in 1997 for travel demand reduction has been replaced in 2000 by renewed interest in capital investment in roads, railways and urban transport.

The implications of this change in policy direction are profound, particularly for the achievement of traffic reduction targets and reductions in emissions levels. The baseline forecast is for traffic volumes in England to increase by 32% (1996–2010), with average time delays per vehicle rising by 11%. If the strongest policy options are introduced then average time delays will fall by 33% (even though traffic nationally will have risen by 21% – 1996–2010), with CO_2 emissions levels also falling by 10% to 27.5 MtC.[4] The strongest policy option consisted of higher levels of road user charging and workplace parking levies in combination with public transport, walking and cycling improvements in *all* urban areas. There would also be targeted charging on interurban roads, and some increases in capacity on these inter-urban roads and high levels of rail investment. Real fuel duty would increase by 19% (2000–2010 – DETR, 2000e). But even with this level of activity, the 20% reduction in CO_2 emissions target set by the Labour government will not be achieved in the transport sector. In 1990, the transport sector in England produced 28.5 MtC with an expected increase of 2.3 MtC to 2010 if there was no major policy intervention. The strongest policy options (stated above) would result in 27.5 MtC or a 4% decrease on 1990 levels (DETR, 2000e). The assumptions used have been criticized (Commission for Integrated Transport, 1999) as being unrealistic with no account being taken of the expected increase in car ownership or the generation effects of new roads.

More fundamentally, there still seems to be no coherent alternative to building more roads. Demand management has offered short-term respite, but unless effective charging is introduced, this will not provide a solution. Even then, if most of the traffic growth takes place outside the urban areas, there seems to be far fewer opportunities for charging, except on the motorway network. The cycle has turned full circle and road building is back on the political agenda with large budgets available for investment. The government has committed itself to increasing expenditure from 0.6% of GDP (2000/01) to 1.8% of GDP (2003/04). It is here that other considerations will again emerge, principally the strong resentment of many to the construction of new roads near to where they live. The public participation process has played a key role in delaying the decision-making process and in making investment in the road infrastructure unattractive to the private sector (Chapter 4). A review of these processes is essential if a coherent roads pro-gramme is to succeed. Otherwise, incrementalism will continue to dominate.

Notes

1   The sequential test is part of the requirements placed on developers to demonstrate that there is no suitable alternative location for their proposal (for example, for shops or housing) within the urban area. Only if this can be demonstrated will planning permission be granted for a greenfield site or peripheral development.
2   Brownfield land is developed land that is no longer needed for its original use – greenfield land is virgin land that has not been previously developed.
3   Urban is defined as over 10,000 people – urban areas cover 7% of the land area, with 80% of the population.
4   The government has moved away from national road traffic reduction targets (related to the volume of road traffic) to indicators such as congestion and pollution. Congestion is measured in terms of vehicle journey times – see Table 5.9.

Chapter 6

The limitations of transport planning

The theoretical arguments

Transport planning has now emerged from a lengthy gestation period, and as maturity is reached it is appropriate to stand back and assess the ways in which the theories and processes have responded to radically different sets of political constraints. This Chapter examines transport analysis over the last four decades from a theoretical perspective and attempts to explain why very little change has taken place in the basic transport planning models (TPM). It presents a retrospective review of the TPM, comments on its limitations, and then identifies the main responses in terms of analytical approaches which were developed in the 1970s and 1980s. Finally, it is argued that since 1990 a renaissance has been taking place because for the first time there has been political and public concern over the way in which transport planning has been carried out. It is no longer acceptable to use the conventional methods which have serious limitations, cannot accommodate the types of changes now taking place in society, and have lost public confidence. Alternative analytical structures are now required and the Chapter ends by suggesting where this renaissance is leading.

Transport planning has developed in parallel with urban planning but seems to have avoided the major theoretical debates which have influenced social science thinking. Two possible explanations might be that transport is immune to such debate or that the unique combination of engineers, economists and social scientists working in transport make it difficult to have such a critical review. The latter explanation may be more likely as most of the criticism has come from outside, but even then it seems to have been primarily aimed at planning in general. Transport planning seems to have escaped.

Table 6.1 attempts to place a general structure on the evolution of three strands of the debate identified by Yiftachel (1989). The view in the 1960s was that planning operated in the public interest, and that it could anticipate the needs of the public and act accordingly. This view was questioned in the 1970s from two basic viewpoints. The Marxists claimed that planning was an arm of the capitalist state and that it was merely facilitating capital accumulation and legitimation of the capitalist system. The contrary Weberian

Table 6.1 The evolution of the debates in planning theory

Decade	Theory	Planning and City Development	Transport	Procedures
1970	Weberian analysis Corporatism Marxist analysis Pluralism	Natural expansion Containment Corridor development	Highway construction Management Public transport and subsidy	Systems analysis Incrementalism Mixed scanning
1980	Managerialism Reformist Marxism Neo-classicalism	Decentralization Renewal Consolidation Sustainability	Market dominance Gridlock	Advocacy Positive discrimination Pragmatism
1990	Company State Social Market	Dispersed cities Cities and technology Compact cities Sustainable development	Road and rail construction Partnership investment	Targets and indicators Financial assessment Social welfare

Source: Based on Yiftachel (1989) and updated to include the 1990s

view was that the state served a multitude of societal interests which were increasingly controlled by a rational and independent bureaucracy. The corporatists argued that there was an alliance between government and industry so that each operated in the interests of each other. Since that time the theoretical arguments have become more disparate until the re-establishment of neo-classical principles and the new concepts of the company state. Here, the market is liberalized, the private sector effectively takes over, and the role of the state is reduced to a facilitating function (Banister, 1990). The interests of society as a whole are of secondary importance to the profits for the company state together with returns to their investors.

Transport has been largely immune to these theoretical debates as there has been a strong state role in terms of public expenditure, regulation and control, and in terms of public transport operations. Throughout the period of ideological debate and questioning, transport seemed to be excluded as it was perceived as a public good to which everybody had a right. The strong social policy tradition made it difficult to place any particular ideological interpretation on transport. However, this has all changed. The origins of theoretical criticisms of transport can be found in the 1960s and the 1970s with the growth in car ownership and road investment, but it really came to life in the 1980s with the radical policies of the Conservative governments.

In transport there has been a tendency for theorists to concentrate on the public/private divide and to relate this to income and the manual/non-manual divide (Dunleavy and Duncan, 1989). The pluralist intellectuals' position would be as protectors and sporadic interventionists, mainly as defensive interest groups protecting the current position. Although there were widespread concerns by these people over the quality of public transport, it was the urban road investment programmes of the 1970s and the public inquiry process which allowed the pluralists fully to articulate their concerns. The pluralists were often ranged against the elitists who represented the vested interests of construction companies in road building and the motoring organizations. The elitists were more interested in the exchange value than broad local use values and they also represented corporatist views (Whitt, 1982). The Marxist sociologists placed transport in the broader sphere of urban social movements by arguing that the basic conflict was between the advocates of the capitalist economy and those who saw broader social priorities in urban development (Lojkine, 1974).

It has only been in the 1980s that the theoretical debates became dominant in transport and other sectors of public sector activity, with the reintroduction of neo-classical economics and its extension to the company state. The notions of a welfare state and public provision of services have been replaced by the market and private sector operations. Transport has been forced to adapt to this political imperative and enter into ideological and theoretical debates, which in the past it had managed to avoid. Previously, the debates on existing transport systems had revolved around different ways of organization in the public sector and whether subsidy should go to the user or the operator. There was virtually no debate about whether organizations should be in the private or the public sector, that was taken as given. Even on questions of regulation there seemed to be some agreement. The pluralists interpreted regulation as the state's willingness to intervene against vested interests of corporate business. The elitists argued that vested interests welcomed intervention as it promoted stability and allowed concentration. Even the Chicago School of economics made a similar argument as they felt industry was likely to control the regulatory agencies itself. This was because the interests of rational bureaucrats and politicians in these agencies get bound up with satisfying pressure from organized interests in their immediate environment. Industry pushes for regulation primarily as a means of limiting competition, stabilizing markets, and promoting collusion between businesses (Stigler, 1975). These different theories have now been tested in practice with extensive regulatory reform and privatization in transport.

The second and third columns of Table 6.1 summarize the changes in city planning and transport over the review period. In both cases the cyclical nature of policy can be identified with growth, expansion and concentration. The major issues in the 1990s were technology and sustainable development, both of which provide new challenges such as the choices between allowing further dispersal of cities (technology) or concentration of cities (sustainable development). This choice reflects a major policy dilemma between ever increasing levels of transport mobility and the political objectives of stabilizing the emission of environmental pollutants. Mobility, cities and lifestyles all have to become compatible if the objective of sustainable development is to be achieved.

As outlined in Chapters 2, 3 and 4, after the golden period of planning in the 1960s, there was a period of re-assessment and critique. It was argued that the systems approach was too ambitious and the theory too simple. In the 1970s the systems approach was also attacked for not being relevant to the policy issues of globalization of the world economies, the geopolitics of oil, and the industrial restructuring which was taking place. The established theories became even more irrelevant in the 1980s with the overt politicization of planning and transport. The ethos of rationality and comprehensiveness, implicit in the systems analysis approach, is impossible to maintain in such a politicized decision-making environment. The 1990s have reflected this explicit politicization of planning and transport as successive governments have tried to downgrade these issues, through reducing levels of investment and through a general lack of radicalism in terms of proposals. But the problems still remain and at some point commitment and action are required as both transport and planning form a integral part of the public services commitments made by government.

Planning analysis seemed to adapt much more readily to this environment (Table 6.1), with diversions away from the systems approach to incrementalism (Lindblom, 1959), mixed scanning (Etzioni, 1967), advocacy planning (Davidoff, 1965), and positive discrimination (Blowers, 1986). Planning procedures were seen as potential instruments for affecting the outcome of political processes, but all these interventions were short lived as the political constraints changed. The only stable paradigm seemed to be the pragmatic rationalization which concedes that absolute rationality and comprehensiveness are not only impracticable but also politically impossible in the increasingly politicized environment of governmental decision-making (Yiftachel, 1989).

Although rational analytical processes should be maintained, it is argued that adaptability and flexibility are the keys to maintain relevance to the 'institutional and political situation in which the planner operates' (Faludi, 1987). This flexible pragmatism has much in common with Alexander's (1984) contingency approach to planning which attempts to synthesize research findings with normative prescriptions. But, in the short term, rational models together with evidence from empirical research and pragmatic experiences are likely to continue in use.

More recently, planning theorists have developed concepts of communicative planning, which involves a more proactive planning process with and by all stakeholders (Healey, 1997). The interest here is that communicative planning takes an explicitly social and institutional perspective on problem solving, as it totally rejects the systems approach. Policy discourses are developed to cover the different policy communities (stakeholders), policy networks (linkages between stakeholders) and policy arenas (institutions where policies are discussed). These elements relate both to the hard infrastructure (for example, formal organizational structures, laws, subsidies and taxation) and the soft infrastructure (for example, social relations, informal networks and professional cultures). It is through combining these two elements that effective change can take place. If only the organizational structures are changed without attention to professional culture and informal routines, little real change will take place (Flyvbjerg, 1998). Such thinking may be at least as important with respect to transport planning, as real change in thinking needs to focus on the soft infrastructure. However, as Voogd (2001) points out, there may also be a paradox in communicative planning, as individual interests have to be reconciled with collective interests. The question here is whether a consensus can ever be reached (Innes, 1995), or whether conflict is an integral part of planning processes. Some of these concepts are now being introduced to transport planning (Vigar, 2001 and Willson, 2001).

In transport analysis there has been no similar process of experimentation with other paradigms. The systems analysis procedures have remained supreme throughout the last 40 years with alternative approaches either being ignored or marginalized. Modification has taken place, but there has been no fundamental reassessment of the basic structure of the transport planning approach which is still consensus seeking and prescriptive. Political economists (for example, Dunleavy and O'Leary, 1987) have criticized the approach and procedures used in transport planning, but with little effect. For example, it is argued that cost utility approaches to evaluation are one of the most distinctive features of technocratic policy planning in transport. These procedures bundle together contestable or implicit social and economic valuations under the guise of sophisticated planning methodology, with easily quantified or monetized aspects of issues being valued over and above the qualitative intangible considerations. This must erode the importance and scope of political debates and processes in decision-making (Self, 1975).

Even the New Right has criticized the cost utility techniques (such as cost benefit analysis) as an insidious tool used by budget-maximizing bureaucrats to justify over-supplying agency outputs at levels which would never be sustained under private market operations (Dunleavy and Duncan, 1989). This thinking has now been supported by the singular lack of interest from the private sector in participating fully in the re-investment in the transport infrastructure.

Despite these forceful criticisms of the technocratic transport planning process from both ends of the political spectrum, the procedures remain intact. There is a realization that many of the decisions are not matters of expertise but matters of opinion, of values rather than facts. In short, decisions are political but this has not made any real difference (Blowers, 1986). Transport planners have tended to remain neutral rather than play a significant role in decision-making. They have tended to scale down their goals and accept the objectives and values set by politicians (Reade, 1987).

This means that they have made a positive choice to remove themselves from having a significant impact on policy formulation and decision-making. Instead, they have cocooned themselves with a commitment to a technocratic role in transport planning and have restricted themselves to the relative comfort of expert advice. By adopting such a low risk strategy their impact has been lessened, but their survival has been assured. This contrasts with the more political view taken in planning where the attractiveness of the politics bandwagon has proved irresistible, and planning analysis has become explicitly political. Ambrose (1986) describes this 'slide' of British planning into the political arena with the role of impartial technical advice being marginalized. The choice is not easy. The limitations of transport and planning procedures are manifest and the decision may be merely whether marginalization of transport planning results from positive action through becoming overtly political, or from inaction through maintaining the *status quo*.

The limitations of the transport planning model

The basic structure of the transport planning model (TPM) has proved robust and long lasting. The four-stage aggregate model, as originally developed in the United States in the 1950s (Chapter 2), has been the bedrock upon which analysis has taken place. Its value has been its ability to examine the city and region at the aggregate level and to establish relationships between a given land-use pattern and travel. The existing situation could be modelled through the four linked submodels (Trip Generation, Trip Distribution, Modal Split, and Trip Assignment), and current traffic problems could then be identified. The modelling process would then be used to predict overall travel demand for the forecast year, and alternative transport strategies would be developed. It was probably the classic example of the systems approach to analysis, and represented the positivistic, descriptive and prescriptive dimensions which epitomized such an approach. The rigidity of the approach has been both a strength and a weakness. Its strength has been in the logic of the process and the representation of the ways in which decisions are made, and it also provided a framework within which transport options could be tested. However that framework has also been a source of criticism in that it has acted as a straitjacket with increasing concerns over the relevance of such an approach to analysis. The TPM is still extensively used in part or in its entirety, not because it is ideal, but because practitioners are comfortable with it and because no adequate replacement has actually been proposed. It is a tribute both to the TPM's robustness and to the inertia within transport analysis that such a situation exists. The limitations of the conventional TPM are well known (Supernak, 1983; Polak, 1987; Atkins, 1987; Supernak and Stevens, 1987), and some of the main points are highlighted here.

The theoretical basis is weak

The TPM is empirically based and designed to predict travel on the basis of establishing relationships that link travel with socio-economic and other variables. This positivistic approach is data driven and makes no attempt at understanding the behaviour of people. As such it is the definitional and measurement issues which become important rather than understanding the reasons why people have to travel. It also explains why much research effort has been directed at improving the methods of analysis rather than the underlying theory of travel. The concept that social behaviour can be explained by physical laws is an attractive one and one with which transport engineers are comfortable. Practical concerns over whether changes can be predicted on the basis of empirical relationships take precedence over the more scientific ideals of understanding travel. Transport is one issue which cuts across theory and practice, with the pragmatists being in the ascendancy over the theorists. Transport analysis has been seen as supporting decisions which have to be taken on investment and subsidy which often involve large sums of public expenditure. Operational approaches have been dominant over the scientific models which may increase our understanding of urban phenomena.

Social scientists in the 1980s tried to redress the balance with considerable research being directed at new approaches to transport analysis (Banister, 1998), but no alternative general theory seems to have been developed and the operational, pragmatic, predictive models are still in almost universal use. The established structure of the TPM has maintained its position for the lack of an acceptable alternative and for the problems that social scientists face when developing alternatives in a volatile social context. Urban and transport modelling are driven by public policy which is inherently unstable and the requirements of decision-makers reflect this instability (Batty, 1989). It may be mistaken to seek an alternative paradigm to replace the existing TPM, even if there was an obvious candidate. More appropriate may be the desire to realize explicitly the limitations of the conventional TPM, and to seek to make improvements whilst at the same time exploring other means by which advice can be given to decision-makers.

A highly structured approach to planning and transport evolved in the 1960s and 1970s which was based on the rational decision model. This process of optimization allowed goals to be set, problems defined and solutions generated by searching across a sample of alternatives characterizing the solution space, with the best plans chosen, then implemented after solutions had been rigorously evaluated against the prior set of goals (Batty, 1989).

The structure of the TPM

Serious criticism has been raised against the structure of the TPM. The sequential structure with the output of the four component parts was seen as having some logic in that the decision to travel could be divided into a series of discrete stages: whether to travel (Trip Generation), where to travel (Trip Distribution), which mode to use (Modal Split), and what route to follow (Trip Assignment). However, decision-making is sometimes simultaneous in that the decision to make a trip includes the subsequent decisions. This argument may be particularly relevant where travel patterns become routinized and habits are formed. The concept that an individual is a rational decision-maker who has full knowledge of all alternatives available (including destinations, modes and routes) may trivialize the actual behavioural decisions being made, including constraints from other household members, previous experiences, linking activities and inspiration. The certainty of human behaviour structured into the four stages and the assumptions on which decision-making are based in no way reflect either patterns of behaviour or uncertainty in behaviour. The concerns in the aggregate studies are over average behaviour, but change mainly takes place at the margin and this variability is ignored.

Within the TPM, the main concern is over testing alternative transport strategies. The interactions between land use and transport are given superficial discussion, with changes in land uses, employment and population being input as exogenous variables. The continual process of change within urban areas is therefore ignored and solutions to congestion are examined solely in terms of transport. The means by which the location of activities can be used to influence levels of transport demand are excluded, and the impact that rises in property prices and commercial rents might have on travel patterns has never been tackled. Similarly, the evaluation of transport alternatives is seen in terms of user benefits, principally the savings in travel time. The more sensitive readjustment of participation in different sets of activities and in activity relocation is not covered, nor is the overall assessment of the implications of the proposal on particular groups in society. The aggregate user benefit means that more and faster travel are the principal aims of investment, and that the quality of travel and accessibility concerns are reduced to a second order of importance.

The structure of the TPM also makes it difficult to include unconventional or radical policy alternatives. It is essentially reactive and constructed so as to produce quantifiable relationships which are assumed to be the key determinants of future demand. Once these relationships have been established, between say socio-economic status and car use, they are assumed to remain stable over time. It then becomes difficult to examine, for example, the impact of technology on car use or area-wide policies such as traffic calming. The structure of the TPM is always promoting more travel and greater levels of mobility. The modelling process is concerned with estimating the scale of the increase and where it will take place. Policies that explore the possibility of reducing travel demand and shortening trip lengths do not seem to fit into such a framework.

Data have always proved to be a strength and a weakness in the positivist approach, as on the one hand the models are data driven, but on the other hand there are never enough data. Most TPMs are static and are calibrated on one set of cross sectional data. Again, assumptions have to be made on whether these data are representative and whether one can predict behaviour from such information. Before and after data and various forms of longitudinal studies allow some of these assumptions to be tested. But the weakness of the primary data inputs to the TPM is not often acknowledged.

Forecasting

Forecasting has proved to be an 'Achilles' heel' of the TPM and it seems that increasing technical sophistication has not resulted in corresponding increases in their accuracy. Part of the explanation may relate to the theoretical and the structural issues, in particular the aggregate scale, the sequential structure, and the data driven nature of the models, but more important than these is the assumption that the future can be predicted with any degree of accuracy. Two crucial issues in all TPMs have been the assumption of stability in model coefficients over time and the assumption that variables excluded from the model will not be instrumental in modifying travel behaviour over time.

Not all significant variables can be specified in the model – there are strong assumptions made on the quantification of variables (principally on the value of time and life, and more recently on the environment), and even the primary unit of study (the trip) is not a clearly defined object. Many types of movement are omitted from the TPM, such as short walk trips, interchange trips and other types linking movements, as the trip is normally defined as the full journey from one origin to one destination rather than the component parts of that trip. For example, the walk to the bus stop, the ride on the bus, the transfer to rail and the walk to the destination are all considered to form one journey. But how many trips, what mode is used, and what are the constraints on that journey? If the person stops to buy a paper or a sandwich on the journey, what happens to all the definitions agreed above? Added to this complexity is human fallibility as recall on trips made is often not accurate.

The use of activities as a framework within which to place trip-making has resulted in better levels of recall (some 15% increase in activities) as all time is accounted for (Jones *et al.*, 1983). Similarly, travel is averaged over the week to find a stable pattern with most effort concentrating on the journey to work. Patterns of travel do vary between days of the week (Banister, 1977), but with much greater flexibility in the use of time that variability has increased with the complexity of lifestyles. Similarly, to focus attention on the work trip oversimplifies travel patterns, as work and commuting trips now account for less than 20% of all trips (DETR, 2000*a*).

The conclusions reached by Evans and Mackinder (1980) on the predictive accuracy of British transport studies were revealing. The average forecast errors over a 10-year period were 12% for population and 13% for employment, whilst car ownership and household income levels were overestimated by 20%. Highway and public transport trips were overestimated by 41% and 32%, indicating that in addition to forecasting errors in the exogenous variables, there were specification errors in TPMs. In general, 'for none of the forecast items would an assumption of no change have yielded markedly larger forecast errors'. Similar conclusions have been reported with respect to US experience (Institute of Transportation Engineers, 1980) and for Europe (European Conference of Ministers of Transport, 1982).

It is not just the inaccuracies in the expected growth in exogenous variables (for example, population and employment), but in the structure of the four-stage modelling process itself. Crucial to the positivistic approach is the notion of the best fit model. The model is calibrated for the existing situation, but errors can infiltrate the process through mis-specification of the base year model, through measurement inaccuracies, and through forecast errors in the exogenous variables. It is essential that in all modelling some form of validation is carried out, either through before and after surveys, or through the checking of model performance against an independent data set.

This point has been returned to in the National Audit Office's investigation into road planning (1988), where again it was acknowledged that traffic forecasting is not an exact science. Possible sources of error included human and measurement errors, statistical errors, modelling errors, prediction errors on specific routes, external factors (such as GDP and changes in fuel prices) and social factors (such as car ownership growth). Taking these factors into account, the Department of Transport considered that reasonable agreement between forecast and actual traffic flows 'can be assumed if their forecast is within +/–20% of the actual flow on the road one year after opening' (National Audit Office, 1988, para 4.1).

The explanations for the errors are not hard to find in the TPM, as the standard procedures used by the Department of Transport only consider trip reassignment when this is just one of a series of readjustments resulting from the construction of a new road. Other changes include travelling at a different time, the use of a different destination, the greater use of the car, and the generation of new trips. Conventional procedures have used a fixed trip matrix and growth factors on flows. The implications are clear, namely that on all fronts there are likely to be differences between forecast and actual trip levels because of mis-specification of the problem.

Even the tests used to check accuracy may be misplaced, as the similarity of predicted and actual flows may have occurred by chance rather than the accuracy of the modelling process. The National Audit Office examined 137 sections of road opened since 1980 and found that in about one-half of the cases predictions were under or over estimated by at least 20%. The question is does it matter? The answer here must be in the affirmative as it influences the decision to build the road, the standard to which that road is built, and the opportunity cost of the capital. The Public Accounts Committee (PAC, 1989) was even more critical and listed seven points which should be addressed by the Department of Transport (Table 6.2). They also commented that the method used by the Department gives a misleadingly favourable view of the under or over estimation of traffic.

In short, there seems to be considerable concern over the TPM and the forecasting process on all fronts. In addition to these theoretical and technical issues, there has been an unprecedented attack on planning as a legitimate process, with the switch away from long-term strategic analysis to shorter-

Table 6.2 The recommendations of the Public Accounts Committee

1 Evaluation is needed of the effectiveness of completed road schemes together with monitoring the out-turn of the original forecasts and appraisal assumptions.

2 Evaluation of the environment is important in reaching decisions on the roads programme.

3 Traffic forecasts must be improved and the Department of Transport's reluctance to accept that there is a serious problem 'verges on complacency'.

4 The M25 is a particular problem and the inaccuracies of the forecasts cannot be fully explained by economic growth. There are also generated and redistributed traffic, and regional factors.

5 Support any moves to speed up the road planning process.

6 The Department of Transport should assess the additional costs to industry, commerce and the Exchequer of not providing extra capacity.

7 The Department of Transport should take account of the wider and more strategic consequences of building new roads.

term market led strategies. This is a move away from the comprehensive city level analysis with the intention of understanding how cities work and the relationships between transport and land use towards much more specific and partial analyses. These changes have been paralleled by the reorganization of local government and the abolition of strategic planning in the metropolitan counties (and London), regulatory reform, privatization, and the primary concern over efficiency in transport and reductions in levels of public expenditure (Chapter 4). With the new social market paradigm and renewed interest over the environment and social welfare, together with the rediscovery of planning (Chapter 5), the TPM again becomes an important component in deciding what infrastructure needs to be built.

Comment

Despite these strong criticisms, 'improvements' have been made to existing approaches rather than shifting to some new paradigm. Such a paradigm shift does not occur through overt criticism of existing practice, as this tends to result in professional barriers being erected and a defensive protection of established practice. To achieve a paradigm shift, the critique must be made in terms which the existing practitioners can accept and relate to. This means that it must be made in terms that are familiar to them, and such arguments relate strongly to the communicative planning paradigm (Healey, 1997).

In transport, such an opportunity has occurred. The TPM was primarily developed to allocate growth in population, in economic activity, in income, and in car ownership. It was concerned with the provision of more transport infrastructure through the expansion of the road network and, more recently, improvements in public transport supply. As Hutchinson succinctly states (1981):

- The provision of large amounts of road capacity encourages the use of cars for commuting, but more importantly encourages long trip lengths between home, work and other activities.
- Public transport demands are influenced by car ownership, but the principal determinants of public transport use are employment concentration on public transport routes combined with high parking charges or parking space constraints on these employment locations.
- The principal dimensions of transport demand, the efficiencies with which these demands can be satisfied, are dictated by factors external to the transport sector such as the housing market, household structure and income, and employment location decisions.
- There is a great uncertainty in the social, economic and technical environment of transport systems, and transport plans must be sufficiently robust and adaptable to changes in the transport environment.

These perceptive comments made over 20 years ago are still relevant today and reflect the in-built resistance to radical change. It has only been with the impact of renewed growth in traffic in the 1980s that some convergence has taken place on the view that increasing supply will not provide a solution. Travel demand overall is increasing at a rate which is substantially faster than effective capacity will (or can) be expanded (Goodwin *et al.*, 1991). At about the same time the Department of Transport (1989*a*) published its *National Road Traffic Forecasts* which confirmed the view that the expected growth in traffic over the next 30 years would be in line with growth in Gross Domestic Product. Traffic would grow by between 83% and 143%. Since then there has been a near universal call for more investment in public transport and restrictions on the use of the car through physical restraint and pricing for road space. In retrospect, these forecasts marked a watershed in thinking as it was accepted that road building would not solve the transport problem and that traffic and demand management would have to form an integral part of any strategy. It was the SACTRA report (1994) that brought two issues to the forefront of the debate. Even though new roads would improve conditions for car drivers, traffic congestion would get worse and that new roads generated additional traffic.

The irony here is that similar views were expressed in the 1960s when a similar growth in traffic was experienced (passenger-kilometres increased by 85% in that decade). These demands were dissipated in the 1970s as economic factors and global recession resulted in lower levels of growth (passenger-kilometres increased by 22%), and other policy concerns absorbed the attention of politicians and the public. They re-emerged in the 1980s as traffic increased by over 40%. Although there was less growth in traffic in the 1990s (only a 7% increase in passenger-kilometres), the economy continued to expand, but with little increase in transport capacity and the consequent increases in levels of congestion on all modes of transport.

The response in the 1970s and 1980s

Research has proceeded along several different dimensions in the attempt to respond to at least some of the shortcomings of the traditional four-stage aggregate approach. In the 1970s and 1980s, there were two main strands of thinking, one concerned with improving the existing approach and the other concerned with the individual choice process.

Integrated land-use transport models (ILUTM)

One of the basic limitations of the conventional TPM was treatment of land use as an exogenously determined factor. The model was primarily concerned with changes in transport demand and in the evaluation of transport alternatives. The ILUTM still took a systems approach to analysis but attempted to model the interactions between land use and transport explicitly within the modelling process, principally through various concepts of accessibility. Different approaches have been adopted. Regression models have been widely used to develop an understanding of the spatial distribution of land uses by making the dependent variable the population, employment or housing in each zone, and using measures of accessibility to determine the appropriate distribution. Different land-use variables can be modelled through sets of simultaneous equations, and the output would be a series of measures relating to the level of activity within each zone. Mathematical programming models allow the allocation of activities to each zone to be optimized. The Australian model TOPAZ (Technique for the Optimal Placement of Activities in Zones) optimizes the allocation of activities to zones by minimizing the total costs of establishment and travel, subject to the constraints that all activities have to be located and all zones are filled (Brotchie *et al.*, 1980). The main determinant of the allocation is accessibility which is included in the form of a gravity model. Similar approaches have been used in the SALOC model developed in Sweden (Lundqvist, 1984). By definition, optimization produces a solution to the specified problem, but makes no attempt to suggest the means by which that solution can be achieved, and no indication is given of the competition process between different land uses, nor to the important role that the pricing mechanism has in determining land use.

By far the greatest effort and interest has been in the use of spatial interaction models, which have strong links with the conventional TPM and the Lowry Model (Lowry, 1964). The model is comprehensive, as it covers housing, residential location, employment, shopping, journeys to work and shops, and land allocation processes. Research in the United States includes the Projective Land Use Model (PLUM) and the Integrated Transport and Land Use Package (ILTUP) originally developed by Goldner (1971), and modified by Putman (1983, 1986).

In Britain, apart from the pioneering research by urban modellers in the 1970s (Chapter 3), most recent attention has come from Echenique and Mackett. MEPLAN (Echenique *et al.*, 1990) is a representation of the spatial economy of a region and it explicitly links the interactions between land uses and transport. In each zone, MEPLAN estimates the equilibrium price of land or building resulting from the demand created by activities in the zone and the supply of land or building at a particular point in time. The equilibrium price of transport is also calculated in the form of accessibility, and this forms the main input to the model for the following time period. Land uses determine the demand for transport and accessibility determines the price and hence the demand for land or buildings.

The LILT model (Leeds Integrated Land Use Transport Model) modifies the basic Lowry structure to include modal split, capacity constraint and car ownership (Mackett, 1985). More important though is the introduction of a semi-dynamic structure to the models with a progression from the base year over a period of time, typically in 5 year periods, with lag mechanisms to link land use and transport at various points in time to ensure that all constraints are met.

The culmination of the research on the ILUTM has been the ISGLUTI (International Study Group on Land Use Transport Interaction), set up by the Transport and Road Research Laboratory in 1981. This comparative study takes seven predictive models and two optimizing models, and carries out a series of comparisons on them (Table 6.3). It was felt that many of the models had similarities and a common theoretical base, with the extensive use of entropy maximization procedures, spatial interaction models (gravity models), and the location of basic employment as the main driving force (Webster and Paulley, 1990).

A range of common land-use and transport policies have been tested on the different models, and some of the models have been calibrated with different data sets to determine the differences in output. The difficulties of international collaboration and the problems of transferability of similar

Table 6.3 Participants in the ISGLUTI study

Country	Organization	Main members	Name of model and year first developed
Australia	Commonwealth Scientific and Industrial Research Organisation (CSIRO)	Brotchie, Sharpe,	TOPAZ 1970
Germany	University of Dortmund	Wegener	DORTMUND 1977
Japan	Universities of Tokyo and Nagoya	Nakamura, Hayashi, Miyamoto	CALUTAS 1978
	University of Kyoto	Amano, Toda, Abe	OSAKA 1981
Greece	University of Thessaloniki	Giannopoulos, Pitsiava	
Netherlands	University of Utrecht	Floor	AMERSFOORT 1976
Sweden	Royal Institute of Technology	Lundqvist	SALOC 1973
UK	University of Leeds	Mackett, Lodwick	LILT 1984
	Marcial Echenique and Partners	Echenique, Flowerdew, Simmonds	MEP 1968 MEPLAN 1985
	Transport and Road Research Laboratory	Webster, Bly, Paulley	
USA	University of Pennsylvania	Putman	ITLUP 1971

Source: Based on Webster and Paulley (1990)

models have created problems, both for the interpretation of policies and for the evaluation of the output (Mackett, 1990). All of the models are access driven as land-use changes respond to changes in accessibility, and transport is then matched to this new distribution. However, with high levels of car ownership and with industry becoming more footloose, the question arises as to whether access is the correct mechanism, as it is interpreted only in a physical sense (that is the generalized cost of reaching the desired location). Other factors such as the availability of land, land prices and the development potential of land are all important in the location and travel decisions, but these are input exogenously.

Some of the limitations of the TPM have been met by the ILUTM. But the complexity of the land development processes, travel decisions and the rapidly changing forms of industry, of population structure, of lifestyles, and of the use of time all contrive to make progress difficult, if not impossible.

Simulation. Simulation provides another means to model the decision process, often at the level of the individual or household. Simulation specifies a set of possible events and the likely outcomes of each event, and these events are then simulated exogenously with the repercussions of each being assessed in terms of flows on the network. It offers the potential for a reasonably comprehensive treatment of micro-level influences and policy impacts, and experience with simulation is becoming more widespread (Simmonds, 2001). The MASTER (Micro-Analytical Simulation of Transport, Employment and Residence) model is one such application (Mackett, 1990). A given set of supply data is input to the model and the various alternatives in the sequences that a household can follow are specified. The model then proceeds to determine the impact of demographic processes, of migration, residential location choice, economic activity and lifecycle, change of job, and finally transport. Not every household or household member passes through all stages in the micro-simulation at each time period, and a wide range of related transport decisions can be analysed, such as the purchase of the car, the use of that car, and the modal choice decision.

Comment

These two developments, the ILUTM and simulation, have advanced the TPM in several important respects. The links between land use and transport have been made explicit, with the realization that relocation of activities will not necessarily lead to more travel but could result in household relocation, moving job, and the substitution of one activity with another. However, by concentrating on the physical factors, the importance of the economic and social dimensions may be reduced. Micro-simulation can capture both the economic and social factors by exploring housing costs, the costs of job relocation and the car acquisition process. Secondly, these models have moved away from the comprehensive claim of the conventional TPM towards examining the interactions between transport and other factors in the decision process. This marks a greater realism in thinking, and it also provides a closer link with the requirements of policy-makers.

The 1980s probably announced, at least unofficially, the final demise of the conventional TPM. That is not to say that it is not still being used, but the concept of an all embracing approach to transport planning has disappeared to be replaced by a more focused approach which takes particular policies or problems for analysis. The strategic approach, involving extensive data collection and modelling, is not seen now as an essential prerequisite to major decisions in the transport sector. Analysis is likely to be more selective using simpler sketch planning approaches to pick out trends. In addition, statistical and limited modelling analysis can be based on a clear statement of policy objectives. The rational comprehensive approach with the systematic consideration and evaluation of transport alternatives has been replaced by short-term broad based analyses carried out at the strategic level and supplemented by detailed local studies. It is here that broader based modelling exercises and simulation studies, such as those outlined in this section, have an important role to play.

Disaggregate behavioural models

To complement the large-scale integrated land-use transport models and the simulation studies, extensive research has also been carried out at the micro-level on the individual choice process. The basic purpose of these behavioural models is to predict the particular aspects of travellers' behaviour most relevant to the planning of transport facilities (Daly, 1981). This approach has arisen out of the realization that aggregate travel phenomena are the result of disaggregate decision-making by a large number of individuals. If the decisions made by each individual can be modelled and understood, then the approach can be called behavioural with a corresponding theoretical foundation. Three basic approaches have been developed, one based on utility theory and the notions of rational choice behaviour, the second uses the principles of psychological choice theory, and the third places travel decisions in the framework of activities.

Disaggregate utility models. Disaggregate utility models evolved from micro-economics and the theory of consumer choice developed in the late 1960s (Lancaster, 1966). Individuals (consumers) act so as to maximize some benefit or utility, with the individual being represented as exercising his or her choice over the full range of available options, limited only by the constraints of time and money. In transport, this choice was normally represented as discrete alternatives between mutually exclusive modes, such as the choice of the bus or the car for the work journey.

This approach epitomizes the rational choice model where individuals have full information on all the alternatives and where they can rank those alternatives according to their own preferences so that the 'right' choice is made. As such it suggests how trips should be made rather than how they are actually made. In its operational form certain simplifications are made, with the random utility model assuming that individual trip-makers weight the various aspects of travel differently, where these weights are assumed to be randomly distributed according to appropriate probability distributions (Hutchinson, 1981). Different probability functions lead to different choice functions, with the most commonly used formulation being the logit choice function where the weights are assumed to be Weibull-distributed. Utility functions are developed for each mode and disutility coefficients estimated so as to maximize the ability of the model to explain the modal choice decisions of individuals (Domencich and McFadden, 1975). Although modal choice has been the main focus of attention, similar approaches have been used to model destination choice, car purchase decisions and car pooling.

Attitudinal models. Attitudinal models shift the focus away from the concepts of utility maximization to those of satisficing behaviour. The individual is making choices in a situation of partial knowledge and when certain thresholds are reached, action will take place. Attitudinal methods are more subjective in their construction and relate to behavioural intentions and consumer choice. Apart from having a modelling role, attitudinal methods have also been used to improve the understanding of individual choice through exploring the links between attitude, preferences and behaviour, and they have provided the means by which attitudinal variables can be included in existing models. Values can be placed on those variables (other than time and cost) considered important in individual choice. These could include measures of comfort and convenience, as well as service reliability and the perceived safety and security of different modes of transport.

Attitudinal methods mark an important change in thinking as it is implicitly accepted that revealed demand may not be the only measure of demand. Almost all analysis takes the trips made as the dependent variable for which explanation is sought. Attitudinal approaches assume that behaviour and travel are wider concepts and an exploration of the decision processes can help explain behavioural intention as well as actual behaviour. The underlying assumption here is that attitudes influence behaviour with an implicit link between an individual's preferences and his or her choice of mode. So if attitudes can be measured, some preference ranking can be obtained from which actual choices and behaviour can be estimated. The difficulty here is that preferences are broader than actual choices and so behavioural intentions may overestimate actual demand.

The weakness in the above argument is the assumed causality in sequence from attitude to preference to choice and behaviour. This oversimplifies the complex psychology of choice. Behaviour also influences attitude and the two elements may interact to maintain consistency between attitude and behaviour (a cognitive dissonance effect). Individuals modify their attitudes to justify what they actually do, so that they have a positive attitude to the car if they use that car. Similarly, there may be some mutual dependence between attitudes and behaviour. Learning theorists argue that through a process of stimulation and response there is reinforcement (positive or negative) which results in routinization of behaviour and the formation of habits (Banister, 1978). Choice is treated as a process and *not* an event with behaviour being modified through adaptation and learning. The overriding concern of most methods over accuracy of output and predictive capability is replaced by a more modest objective of testing different behavioural assumptions and understanding behaviour. The concern is more with diagnostics than prediction.

However, while attitudinal methods do provide a range of approaches to measure the strength of beliefs and the likely behavioural response to change, their real contribution will only be apparent after extensive applications, and this has now been achieved in transport, primarily through the use of stated preference methods. Whether compensatory or non-compensatory structures are used, testing and validation are the only means by which the strengths of attitudinal methods can be assessed.

Activity based approaches.   Activity based approaches present an important theoretical advance as they view travel as a derived demand rather than a direct demand. This means that people only travel because of the benefits that they perceive at their destination, and that little travel is undertaken for its own purpose. Following on from that premise, the activity based approaches assume that travel is one of a set of activities which people participate in and it provides the link between activities which do not take place at a particular location.

Much of the pioneering work was carried out in Sweden (Hagerstrand, 1970) where a general framework for human interaction has been developed through which travel behaviour can be analysed. The aims of this framework are to demonstrate the activity opportunities and constraints which face particular individuals or groups of people. Activities are constrained by their location (spatially), by their opening hours (temporally) and by whether they can be visited as part of a 'tour' (complementarity). People are constrained by their physical and economic conditions (personal constraints), by their work and school hours (time constraints), and by linkages with members of their own family and others (coupling constraints). In addition to these two sets of limitations there are also authority constraints which are imposed from out-side (for example, opening hours for shops and bus timetables). Travel is assumed to be limited by these constraints and so the space time prisms can be constructed within which any individual's opportunity set lies. It is here that transport plays a crucial role as the shape and size of this prism is often controlled by the availability of a car. In essence, such a constraints-based approach is a physical representation of society which explicitly portrays travel and non-travel activities. Within this framework the dynamics of behaviour can be investigated, and it allows an understanding of the types of specific adjustments which might take place as a result of a policy change (Carlstein *et al.*, 1978).

Activity based approaches have been extensively used in a variety of specific applications:

♦   Diary studies have explored the use of time and space over time, in particular locations, by different social groups, and subject to particular constraints (for example, car availability). A recent development is the episode-based activity survey formats to understand in-home and out-of-home activities (Pas and Harvey, 1997).

♦   Time geographic accessibility studies have combined all the concepts into a modelling approach based on two-dimensional space and time. A simulation model (PESASP) provides the answer to the question as to whether an individual's activity path is physically compatible with the constraints imposed, and it shows the number of possible combinations and the different sequences by which activity programmes can be achieved (Lenntorp, 1981).

♦   Interactive gaming approaches simulate travel and other aspects of behaviour over particular periods of time. These approaches can either examine the responses of families to particular changes in policy or focus on selected issues such as the environment and the household budget allocation decision. In each case the respondents, usually members of the household, present their current set of activities, prior to a change being introduced followed by a discussion and the resolution of the new set of activities (Dix, 1981; Jones *et al.*, 1983). These methods have also been used to understand the interactions within households and joint activities, including the trade-offs made (Golob, 1999 and Golob and McNally, 1997).

♦   Activity scheduling models include an explicit dynamic element and examine the linking of sequences of activities through trip chaining or the ways in which household schedules are formulated. In all of the scheduling approaches, time budgets and the use of time are considered to be the driving force in terms of the sequences and the range of activities in which households can participate. The complexity of activity scheduling makes it difficult to model, but new data analysis and simulation methods may help here (Ettema and Timmermans, 1997; Axhausen, 2002).

Activity based approaches are now seen as being in the mainstream of methods available for transport analysis, as richer data sets have been collected and as links have been made with the network supply implications of the predicted travel demand. The focus on the availability and use of time is fundamentally different from other approaches in travel demand analysis. Time is seen as the principal constraint and travel is just one of the many activities undertaken during the day. The research in Portland, Oregon is one example of where activity analysis has been successfully combined with conventional descriptions of transport networks through micro-simulation models (Simmonds, 2001).

Comment

The disaggregate behavioural models are much more modest in scope than the conventional TPM and they are based on different views of how individuals should or do make decisions. In this sense, they are behavioural, but a general theory of travel behaviour has not been developed. Each approach sets out a series of assumptions which allow an analysis to take place, but reality is a mixture of three facets. Individuals are not utility maximizers, nor preference satisficers, nor completely limited in their activities by the constraints of time and space, but a combination of all three.

Where these approaches made a significant contribution was in the extended range of policy options which could be examined, and in the move away from the obsession with long-term prediction towards a short-term view of reactions to particular policy changes. Their concern was over the impacts on particular social groups and it was accepted that travel decisions were complex. However, these methods have not gained widespread acceptance in practice. Heggie (1978) argued that the responses suggested were at odds with empirical evidence on behavioural responses to changes in fares and levels of service, and that there was still an underlying ethos which wanted a statistical explanation rather than a behavioural one.

Despite these limitations, several important policy issues could be addressed and have been analysed by this first generation of behavioural models (1970–1990). These were summarized in Hutchinson (1981):

- changes in family structures and lifestyles and the related impacts on transport and housing markets;
- the ageing of the population and the impacts on transport needs as they relate to other markets such as housing, health services and recreational opportunities;
- the changing employment bases of urban areas and the implications for fixed route, centrally-focused transport facilities;
- increasing energy costs and the impacts on residential, job and retail markets and on transport demands and needs;
- the reductions in economic activities and the ability of governments to continue to finance the deficits of public transport systems;
- the impacts of the expected innovations in computer communications technology on household and institutional activity patterns and organizations.

Available modelling techniques were not capable of responding completely to these issues and observed behaviour might not be the most appropriate way to investigate likely future response. As the uncertainty, complexity and flexibility of individual actions have greatly increased, new methods must be developed to investigate these impacts.

The response since 1990

In the 1960s when urban planning and transport analysis were in their infancy, a series of landmark conferences was held, principally in the United States (Chapters 2 and 3). In the mid-1980s a second series of important conferences was also held to mark the switch in transport planning away from the primary concern over the rational comprehensive approach towards the use of a much wider range of analysis tools. Conferences were held in the UK, Japan, Australia, France, Canada, as well as the US. The output from these conferences was a restatement of the current position with respect to travel demand forecasting and transport planning analysis. However, the more important output has been the attempt to sketch out the new agenda and requirements for analysis.

Goodwin (1991) presented a clear summary of the usefulness of the different models that made up 'the family of approaches':

- Assignment models were used to predict traffic flows for trunk and local schemes. These models assumed that a scheme will only influence route choice within a fixed total number of trips. This meant that travel demand was only marginally responsive to transport supply or costs, and models were calibrated in order to reproduce an approximately known level of flows in a base period, using traffic counts.

- Four-stage models were used for big complicated schemes in large urban areas. These models allowed destination, mode and route choices to be influenced by transport supply, but usually not the frequency or structure of trip-making, (except for committed land-use developments generating a pro-rata increase in trips) or in time of day of travel. They were calibrated in order to reproduce approximately measured patterns of travel on a notional single ('average') day in base year, usually using household surveys and counts.

- Aggregate procedures, using implicit or explicit elasticities or other behavioural parameters, were used to forecast national traffic levels and aggregate public transport usage. They were typically based either on a non-lagged regression of a statistical time series over several years, or discrete choice modelling more or less similar to logit models of mode choice, or sometimes *ad hoc* interpretations of miscellaneous data to hand.

- Surveys sometimes used stated preference techniques to forecast usage of new facilities or service modifications.

- A subtle process of judgement designed to adjust, constrain or 'reinterpret' model forecasts especially when in the initial stages of a forecasting exercise there were odd or counter-intuitive results. This judgement was a necessary part of any forecasting process. It should be remembered however that the fundamental credibility of a model, and the insidious influence of previous modelling experiences in actually forming the intuitive basis on which the judgement is built, still needed to be consciously examined.

The struggle faced by many researchers was that even though they had considerable reservations about the established procedures and there was no lack of new ideas, little change took place. It seemed that research which did not easily fit into the established order was marginalized. Thus the opportunity for improvement or for diversification was rejected in favour of the traditional approaches to analysis. Goodwin (1991) has strongly argued that much more research has been done than is appreciated, and that although it was of direct relevance to the policy needs of the time, it was ignored because it did not fit into the established framework. He goes on to suggest that much of this 'marginal' research should be reclaimed.

Transport planners have been resilient in avoiding many of the theoretical arguments presented by political economists and others in the 1970s, and their reaction was to seek refuge in more technical arguments with which they were familiar. Consequently, the basic framework for transport analysis has remained unchanged, and it has only been recently that greater flexibility has been apparent. In addition to the two strands of research outlined in the previous section, five other more recent developments have taken place.

Together these new developments can be seen as a renaissance of transport planning, as they each reflect new techniques and ideas which have made the transition from research to practice.

Integrated transport studies (ITS)

Integrated transport studies (ITS) have been carried out in many cities in Britain as a base for guiding transport strategies over a 20 year period. Their origins derived from the new requirements of the Department of the Environment (1989) for clear statements on transport policy in the unitary development plans, particularly when a case was being presented for infrastructure investment and new light rail systems.

The procedures required a statement of vision for the town or city which covered economic, environmental, and quality of life aspects, together with the specific transport objectives to meet the vision and a realistic statement on the likely finance available (May, 1991). Against these objectives, potential transport problems were identified and this stage required a prediction of future conditions, but acknowledgement of the inherent uncertainty that existed. Strategies which cover both land-use and transport options were then specified and evaluated against the full range of objectives. The best elements were then combined in a preferred strategy.

The basic differences between the ITS and the TPM were that the time horizon was reduced to the medium term, objectives were very clearly stated at the beginning of the study, and a wide range of strategies was tested. The extended range of objectives might include efficiency in the use of resources, encouragement of economic regeneration, accessibility within the city, environmental quality, and sustainability. The whole study would take between 3 and 6 months, and there was no requirement for detailed model building, data collection and calibration. Strategic sketch planning models were not new but they suddenly came back into fashion. The structure of the Birmingham Study (Figure 6.1) has been used as a role model for other studies (Jones, et al., 1990). The preferred strategy is fully costed and evaluated through multi-criteria analysis, which relied on both model output in terms of meeting expected demand and professional judgement.

The ITS rapidly gained respectability and their use has increased as the approach is familiar to both transport planners and politicians. The basic question is one of balance, whether the ITS is too superficial or whether it does encompass the main alternatives and investigate them to a sufficient depth so that full consideration is given to uncertainty. Indications seem to suggest that they have an important role to play in testing a wider range of options in a systematic and cost efficient way, but that they are less appropriate when innovative schemes are being tested

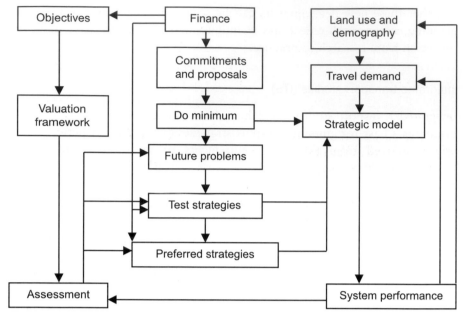

Figure 6.1 The structure of the Birmingham study (Source: Jones et al., 1990)

Evaluation methods

Evaluation methods provide a second area where a radical rethink has taken place, partly as a result of concern over existing economic based approaches and partly as a result of the growing importance of externalities, principally the environment. These issues have been fully discussed in Chapter 3, and here only some issues and examples are given.

The three main concerns being addressed are the comparability between evaluation for different modes of transport, the value of time (which accounts for about 85% of user benefits in road evaluation), and the question of comprehensiveness in evaluation (which includes externalities). The connection between strategic environmental issues and strategic transport planning is poorly developed both analytically and in terms of its presentation to decision-makers (Wood, 1992). A crucial issue here is the appropriate time for the appraisal so that it fits in with the decision process and when the expected impacts might be apparent.

The role of cost benefit analysis in evaluation has been qualified, as other multi-criteria analysis methods (such as the Goals Achievements Matrix and the New Approach to the Appraisal (NATA) of trunk roads projects) seem to provide a framework within which the economic, political and environmental factors can all be placed. Local authorities have made considerable use of Goals Achievements Matrix approaches to evaluation (for example London

Strategic Policy Unit, 1986) but the most comprehensive use of a multicriteria framework was in the approach adopted in the London Assessment Studies (Coombe *et al.*, 1990). The framework identified four main groups of impacts (and two supplementary groups), and presents a large table with one column for each option and rows for each impact.

Group 1   Economic effects on travellers and transport operators including efficiency, safety and costs (travel time and vehicle operating cost changes).

Group 2   Economic costs of implementation, enforcement, maintenance and operation (includes costs of road works, landscaping, land and property).

Group 3   Effects on the human environment (includes users and occupiers of buildings, pedestrians and cyclists).

Group 4   Effects on the physical environment (through brief descriptive statements on heritage and urban structure).

Group 5   Effects of the options on the Boroughs' local objectives.

Group 6   Special effects such as impacts on ecology or agriculture.

The options, which were evaluated in each of the four corridor studies in London (West London, East London, South London and the South Circular), were all placed in this evaluation framework and assessment was based on the achievement of objectives, the effects on problems, and value for money. The options involving public transport were as far as possible assessed on the same basis as the road options.

The NATA projects (DETR, 1998*a*) place the economic costs and benefits as derived from the cost benefit analysis (CBA) in a much wider and more transparent framework. Five criteria form the basis of the analysis. In addition to the conventional economic criteria, there are criteria relating to environmental impact (noise, local air quality, landscape, biodiversity, heritage and water), safety, integration, and accessibility (accessibility to public transport, community severance and pedestrians/others). All the evidence is presented to the decision-maker in a single Appraisal Summary Table (AST), with information from the COBA, the EIA, and the relevant government regional office (Price, 1999). The AST presents the net effects of the road on each of the five criteria, with no weightings allocated to each part. The weighting process is left to the decision-maker. This approach has now been extended to all modes of transport and to the current multi-modal studies. Evaluation has moved away from the single indicator based on rates of return to looking at the broader issues relevant to investment decisions, so that situations can be identified where the costs and benefits related to the other criteria are sufficiently high to outweigh the conventional economic costs and benefits.

Stated preference and contingency valuation methods

Stated preference and contingency valuation methods are two techniques which have been incorporated into standard practice. Using interview techniques and attitudinal questionnaires it is possible to elicit preferences for alternative choices. The process can be carried out either in a non-compensatory structure where the alternatives are eliminated after an attribute by attribute comparison (for example Tversky's Elimination by Aspects Model) or in a compensatory structure which requires a trading off of higher levels of satisfaction on one attribute against lower levels on others, unless the cost is increased (Kroes and Sheldon, 1988; Louvière, 1988). These methods have been widely used to assess preferences for different modes, and to place values on quality of service variables. They are particularly useful where new or radically different proposals are being assessed, and where past trends are unlikely to give a real estimate of future demand.

The contingent valuation method (CVM) goes one step further and actually asks respondents to state a price that they would be willing to pay for a particular alternative, or what they would be willing to receive by way of compensation to tolerate an increase in costs. Again, CVM methods may provide the only choice of method in particular situations, such as valuing the environment. The problems are similar to stated preference in that there is not a direct correspondence between the hypothetical market and the real market, and there is no empirical symmetry between willingness to pay (WTP) and willingness to accept (WTA). Conventional economic theory suggests that an individual should show indifference between WTP and WTA. Preferences and contingent valuation both tend to overestimate demand, and this limitation means that validation is important. Nevertheless, CVM is one of the only methods available for exploring the different means to assign monetary values to the environment.

> Many costs and benefits are measured directly in money terms: for example, savings in expenditure on resources, and sales revenue. Where they are not (e.g. travel time saved, noise and other forms of pollution, and broad managerial or political factors) costs and benefits can sometimes still sensibly be given money values, often by analysing people's actual behaviour and declared or revealed preferences. These imputed money values can be used in the appraisal as if they were actual cash flows. Other factors which cannot be valued should be listed, and quantified as far as practical, making it clear that they are additional factors to be taken into account. It can sometimes be helpful to calculate the value which such factors would have to take for the net present value of a scheme to turn positive or negative . . .

> Account sometimes needs to be taken of the value which individuals may place on the possibility of using a service or visiting an attractive area even when they do not currently use the service or make visits. (UK Treasury, 1984, paras 33 and 34)

These statements suggest that actual behaviour and revealed preferences can be augmented by surrogate markets (for example, hedonic prices) and hypothetical markets (for example, contingent valuation), as well as option value. Again, the crucial transition of concepts from research to practice has taken place, and a broader interpretation of preferences and values can be included in selecting alternatives.

Dynamic analysis

Dynamic analysis has become the focus of substantial research in the social sciences, even though earlier attempts at longitudinal data analysis were inconclusive (Wrigley, 1986). This may have been because the earlier efforts were directed towards the collection of data rather than its analysis. More recently, repeated cross section data sets have become available so that the output from static analysis can be verified and this in turn has led to dynamic modelling becoming of interest to transport researchers. Richer data sets involving diaries and panel surveys have also allowed the dynamics of travel behaviour to be explored (Jones, 1990). The practical use of these data sets and the associated analytical research has been more apparent in the Netherlands and the United States than in the UK (Goodwin and Layzell, 1985).

The theoretical base for such an approach is strong. Individuals do not make decisions in isolation, and subsequent decisions are always influenced by prior decisions. An individual's experimentation, learning and experience all accumulate over time, leading in turn to preferences, values, habitual behaviour, and decision rules. In addition to the dynamics of behaviour there is also the time dynamic. Behaviour is not replicated each day and activities are scheduled over longer periods of time. Dynamic travel analysis investigated both the behavioural and temporal aspects of travel decisions.

Most dynamic analysis in transport has focused on the modelling of specific aspects of the travel decision (such as route choice and departure time), and on establishing a consistent analytical basis. There are methodological problems raised by the type of data collected (mainly qualitative) and the inter-relationships between variables. For example, in panel analysis much research has focused on the difficulties in obtaining and maintaining a consistent analytical framework over the panel waves. As Kitamura (1990) suggests, consideration must be given to:

- possible increases in non-responses due to the fact that respondents are required to participate in more than one survey;
- problems of sample attrition;
- problems of locating respondents in multiple survey waves due to residential location and dissolution of households;
- possible decline in reporting accuracy due to 'panel fatigue';

♦   problems of 'panel conditioning' where behaviour and responses in later
     surveys are influenced by the fact of participating in the panel, and also
     by responses to previous surveys.

These problems are not unique to panel surveys, but are a price which has
to be paid for better quality data. Dynamic analysis has now come of age
(Axhausen, 2002; Stopher and Lee-Gosselin, 1997). Although it may offer
many theoretical and methodological advantages, panel surveys are probably
the least likely of those presented here to be widely accepted and used because
of the additional costs of data collection and the skills required for analysis.

Large-scale land use and transport demand studies

In all the developments itemized in this Chapter there has been one underlying
theme, namely data. Information about travel decisions, about the city, about
demographic factors, whether collected at one point in time or over a series of
points in time, has featured at all stages. Good quality data are difficult and
expensive to collect, and one feature of the 1980s and 1990s has been the
absence of the large-scale travel demand surveys which characterized the
classic land use transport studies in the 1960s and 1970s.

Updating of databases has taken place, but there are now increasing
requirements for new information for monitoring and system performance
measurement. Technology now permits the collection of huge amounts of data
on the use made of the transport system, both by cars in congested urban
conditions and by public transport through the use made of passenger
information systems, automatic ticketing, and smart cards. However, all these
data only report on actual travel behaviour and on the performance of the
transport system. They do not explore the motivations for travel, household
decision-making or suppressed demand.

The concept of travel data as an information source is an important one.
Modelling in the traditional transport planning sense is no longer the main
reason for collecting travel data and the data themselves are not suitable for
input to the TPM. Database enquiries, the use of real time information, and
interactive management systems are all more important concerns for transport
operators and managers. The question then becomes whether this information
can be used to investigate travel behaviour or what limited supplementary
data are required to link the travel data collected so that modelling can take
place. The survey process changes to one of database assembly and
management together with the most appropriate means to interrogate that
information. Surveys will only be undertaken to provide essential
supplementary data and to update the data. All information will link in with
the basic database structure (Taylor *et al.*, 1992). The ready availability of
data means that a new generation of models have been developed to take

advantage of these sources of digital information, and some of these land-use modelling packages are summarized in Table 6.4.

Geographic information systems (GIS) provide the basic structure for many databases as their spatial organization allows information to be added or extracted very easily. Through an interactive graphics interface, it is also possible to 'see' the data. The analyst can also impose a particular structure on the data (for example, household, zone or area) to suit his or her own requirements. This permits analysis at different scales and for linkages between the scales. The database has not been assembled for one purpose but for all users, and so this should allow for a richer analysis of interactions between sectors, as can be seen from the example given below.

Table 6.4 Land-use modelling packages available in the UK

Package and and supplier	Content requirements	Data processing	Data	Calibration
DSCMOD David Simmonds Consultancy	Static, land use, based on accessibility changes and measures of economic potential	Mainly standard census data	Minimal	Limited
DELTA David Simmonds Consultancy	Dynamic, land use, with segmentation of activities to cover decision to move and location choice	Social data, plus information on transport and land uses	Significant and broad based in its approach	Draws substantially on previous research
MEPLAN Marcial Echenique & Partners	Dynamic multi-modal land-use and transport model, based on input-output analysis, a logit model for mode choice and residential location models	Substantial information on activities, travel and location	Significant knowledge and time to set up	Significant
MENTOR Marcial Echenique & Partners	Dynamic land-use model to be interfaced with transport models – distribution of transport demand is derived from the interactions in the land use model	Mainly standard census data	Minimal	Mainly automated
TRANUS Rickaby Thompson & Associates	Dynamic land-use and multi-modal transport model	Substantial information on activities, travel and location	Significant modelling knowledge and set up time needed	Significant
ESTEEM Bartlett School of Planning UCL	Dynamic land-use and transport model, based on existing GIS software	Broad, but easily available sources that are digitally coded	Some database skills required, but modelling interface user friendly	Needs local travel diary

Source: Based on David Simmonds Consultancy (1998)

The ESTEEM[1] model is a good example of one such GIS approach which is designed to assesses the sustainability of new housing developments, together with the associated distribution of services and facilities, by calculating total personal travel demand by mode and purpose (over 80% of all travel is modelled), the use of energy in travel and the levels of emissions for seven pollutants (Titheridge *et al.*, 1999). This model has been tested for three English counties (Leicestershire, West Sussex and Kent) where an additional 10,000 households were located according to four different development strategies – intensification, extensification, decentralization, and new towns. As such, the model has been designed to help local authorities in their development plan review processes, particularly in assessing the implications of integrated policies. The model has key differences from other transport models currently on the market, as it has been designed to use readily available data sets and to link with the ArcView GIS system, which is one of the more popular GIS systems currently in use within local authorities. Thus, transfer of data to the model is minimized (Figure 6.2) as primary data are really only needed for calibration purposes. Even here there are default options. Considerable effort has been invested in the user interface so that the operation of the model is intuitive. This allows users to test different options 'on line' so that the implications of alternative land-use and development strategies can be assessed in terms of their transport outcomes.

Comment

One of the main problems with technical forecasts and trend analysis is the obsession with numbers, not with the underlying principles and ideas. Little thought has been given to visions of the city as a place in which to live and work, and how such a goal can be achieved. Planning and transport planning in particular have both become calculative. As Schon (1983, p. 39) puts it:

> attention to problem solving has been at the expense of problem setting – students have been taught how to put real-world problems into the molds of techniques which appear to render them soluble, but have not been taught how to inquire into whether they are asking the right questions in the first place.

There is a yearning after scientific rigour based on a premise of technical rationality which favours harder physical science rather than softer social science.

The attractiveness of the apparent ease with which quasi-scientific approaches can provide easy answers to difficult questions is understandable. There is a reassurance in numbers and an intolerance of uncertainty. Technocratic analysis such as that found in the conventional TPM gives that reassurance by assuming that travel demand at the aggregate level is repetitive and predictable.

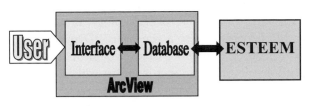

Figure 6.2.1 Model structure

Modelling Strategy

ESTEEM allows the changes in travel, energy and emissions between different strategies and the base model to be plotted as a surface. Figure 6.2.3 shows an example of change in mean trip for Leicestershire. The effect of strategies on energy consumption and emissions can also be plotted.

The Model

ESTEEM consists of a travel and emissions module operating as an extension to ESRI's ArcView GIS package (Fig. 6.2.1). An origin-constrained gravity model is used to simulate travel patterns. The emissions module takes into account fleet characteristics, as well as estimates of cold start distances. The module also includes algorithms for calculating emissions from modes such as electric, CNG and LPG vehicles. ESTEEM allows the user to view the results as tables, charts or maps (Figure 6.2.2).

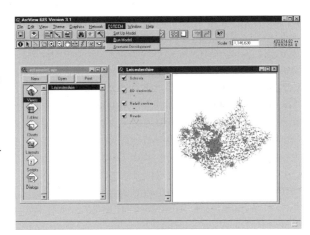

Figure 6.2.2 ArcView project with ESTEEM customized interface

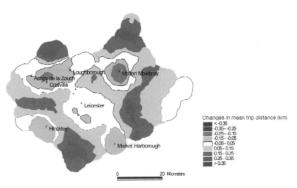

Figure 6.2.3 Changes in modelled mean trip length – just one of the many possible outputs from ESTEEM

Benefits

ESTEEM allows strategic planners to quickly test a large number of possible options for new developments, at a very early stage in the structure plan and unitary development plan review processes. By providing a model which uses a familiar windows based interface, together with the means to link it to their own local data, some of the questions about which settlement options are more or less sustainable in transport terms can be answered.

Figure 6.2 Details of the ESTEEM GIS model and its interface with an example from Leicestershire

But transport analysis does not really explore the reasons for change which relate to economic development, social policy, the globalization of economies, the increase in affluence and leisure, and the use of technology. In one sense, the 1989 National Road Traffic Forecasts were a watershed (Department of Transport, 1989a), with the realization that more roads would not solve the problem, but the real problem is one of single sector vision which does not draw linkages between transport and what is happening in society as a whole (Table 6.5). The case is not to dispense with quantitative methods, but the myth that they are unbiased arbiters in prescribing proper planning actions. 'They must be moved away from centre stage and replaced by a broader more critically oriented approach' (Richmond, 1990, p. 52).

Renaissance in the 1990s and 2000s?

Almost all analysis in transport planning has been based on modelling approaches which crudely represent the economic processes in terms of the demand for movement and the available supply of infrastructure and modes. The links between the models and the actual structure of activities in cities and regions are tentative, and a direct correspondence is assumed to exist between land use and travel. The treatment of travel is as a physical concept. These models, both transport and land use, represent a fusion of gravitational concepts underpinning spatial interaction with macro-economic theory as reflected in input-output and economic base analysis (Batty, 1989). Many of the subsequent developments outlined here have still only been used as applied research tools rather than as policy-making instruments.

There has been little interest from researchers to study substantive new issues and new policy questions from a modelling perspective. As Batty (1989) comments, modelling has acquired a life of its own and has become institutionalized. It has retreated from the volatility of public policy-making into its own cocoon and it has not responded to the challenge of social change.

Table 6.5 New directions in transport planning

1 Move away from trend-based extrapolation to richer social analysis based on linking transport to what people do and how industries operate.

2 Change the balance in evaluation so that greater weight is attached to qualitative factors and ecological arguments.

3 Move away from 'objective' factors in analysis (for example, cost and time) towards the acceptance of subjective valuation and political rationality. The primary concern over time saving should be replaced with measures of improvement in the quality of life.

4 Greater realism in understanding people's rationality in what they do, particularly in developed high-income economies, where the marginal utility of income may be limited.

5 Abandonment of single sector analysis in favour of integrative studies which relate to the city or locality as a whole

Batty explains this essentially defensive strategy in terms of the eclectic set of disciplines from which urban and transport planners emerged, and from the small number of researchers involved in the field. Despite these limitations, the research output has been impressive, at least until the mid-1980s, although the greatest progress has been made in analysis methods rather than as an input to the public policy debates.

Transport planning has been on the defensive as priorities moved from planning growth to the management of decline in urban areas, and away from providing for the car to managing the use of the car. Transport is no longer seen as a means to distribute growth in the urban area, but as a crucial lever in regenerating economic growth in areas where the traditional economic base has been ravaged in the transition to a post-industrial urban society. The critiques of urban planning and transport raised in the 1970s (for example Sayer, 1976) which attacked the positivist approach as being unidirectional in causation and not involving dynamic processes and feedback, now seem irrelevant to the agenda which has evolved. Traditionally, competition was seen in terms of substitution. For example, a new transport mode competes directly with existing modes for patronage. New forms of communications, information transfer, and the restructuring of industry result in increasing linkages between alternatives and additional travel being generated by a wide range of modes. Processes have become much more complex, and the realization that the action-reaction cycle has many new dimensions forms one of the main challenges of the new century.

The questions raised are fundamental. It is unclear whether models in their traditional sense can accommodate the ideological culture of self-interest which characterized the 1980s, and the technological and demographic changes which are taking place in society. The representation of urban systems in spatial terms may only present part of the picture as technological and demographic change are essentially aspatial, but have spatial outcomes in terms of distribution. The culture and attitudes of people, business, industry, and governments all influence what actually happens. All forms of planning have been exposed to the new competitive requirements in terms of their own structure, their accountability, and their range of activities. Not only have the questions changed, but so have the answers.

Space is still a crucial determinant of cities and travel, and it has a fundamental role in exploring planning policies. But it is now only one element in the full range of factors which have to be considered. In his discussion of the 'American city theoretical', Hall (1989) concluded that planning theory had moved away from what planners actually do and the kinds of issues they confront in their working lives. The impasse has to be resolved. Some theorists have moved away from urban planning to political economy (for example, Scott, 1969 and Castells, 1983), whilst others (for example, Markusan, 1985 and Bluestone and Harrison, 1982) have proposed linking political economy approaches with active regional policy prescriptions for older industrial regions. Another group (for example, Friedmann, 1973 and Forester, 1980) have maintained the more traditional link between theory and action. Apart from trying to reconcile theory with practice and formulating an appropriate role for planning, the grass roots movement is returning to hands-on planning to cover economic redevelopment, project finance, and affordable housing. A further group has rejected the positivistic approaches and concentrated on the institutional and organizational issues as they relate to communicative theory, taking account of both formal and informal decision-making structures (for example, Innes, 1995; Healey, 1997; Vigar, 2001 and Willson, 2001).

The dilemma facing transport planners is that even if it is accepted that the methods available are no longer appropriate, that the issues to be tackled are more complex, and that the policy context has fundamentally changed over the last decade, the problems still remain and important decisions, particularly on investment and on large scale projects, have to be made. Substantial investment is needed in the new transport infrastructure, to cover road, rail and air. The scale of that investment is unprecedented as extensive renewal is required and as networks become fully international. These transport investments will be linked in with other projects on a grand scale. Included here are Paris Eurodisney, London's Eastern Corridor, a new high speed rail network in Europe, and East-West development in Europe. Investment will be a combination of private and public capital with such issues as investment appraisal, joint venture funding, and ecological impacts all becoming key factors in the analysis. The traditional single-centred or poly-nucleated city will also change as technologically led deconcentration takes place (Table 6.1), and as the work ethos is replaced by one based on increased leisure time. New pressures will emerge as the centripetal forces are matched by strong centrifugal forces to maintain the city as the only solution to sustainable development and the maintenance of a high quality of life in the densely populated countries of Europe.

From the viewpoint of transport planning analysis (and of urban planning), the need for a renaissance at both the strategic and local levels is required. Three basic reasons can be argued for such a renaissance. Even after a pragmatically led decade where political and ideological concerns were high, some form of strategic framework is necessary. But it is likely to be very different from that advocated in the 1970s and 1980s which was based on the production of large-scale proposals to meet expected shortfalls in capacity. Various forms of strategic and regional guidance would be maintained, irrespective of whether county councils are to be abolished in Britain and replaced by regional assemblies. Within the existing or modified framework for planning, transport planning would perhaps adopt a more flexible form of contingency approach (Alexander, 1984), where empirical research findings are linked more closely to normative prescriptions. Such an approach would allow new ideas, research methodologies, and evidence from research in Europe and elsewhere to be fed into a strategic framework.

Whether investment is public or private sector funded, some stability is required in the decision-making framework so that decisions can be made with some certainty, otherwise only low risk strategies will be adopted. This means that in transport, only those projects which are underwritten by the government would actually proceed. If the private sector is to be fully involved in the new generation of investment, then a planning framework is required and analysis methods are also essential, together with the ability to learn from past experience and from overseas experience.

The second argument is that predicted increases in travel demand over the next two decades cannot be met. The expectation that new capacity could be provided, either through new construction or through traffic management, to meet that growth has now been firmly rejected. Even if it were possible, it has been seen as undesirable. This new realism (Goodwin *et al.*, 1991) has focused the minds of all those involved in transport planning. The principle of the new consensus is this recognition that there is no possibility of increasing road supply to a level which approaches the forecast increases in traffic.

It logically follows that:

♦ Whatever road construction policy is followed, the amount of traffic per unit of road will increase, not reduce; i.e. all available road construction policies only differ in the speed at which congestion gets worse, either in its intensity or in spread.

♦ Therefore demand management will force itself to centre stage as the essential feature of future transport strategy, independently of ideological or political stance (Goodwin *et al.*, 1991, p. 111).

For the first time, even though transport has been a reasonably high priority on the political agenda, there is now a consensus as to what should be done politically and professionally. This means that radical action could take place provided that public support can be obtained, and the resulting requirements for analysis could break the traditional mould of transport planning. However, this apparent agreement has already been modified as the *Transport 2010* document (DETR, 2000*b*) promises a new generation of road building, even though much of the investment will be in upgrading existing roads and in building bypasses. The crucial balance between the different interest groups in transport and in making the best use of available forms of transport has again been readjusted.

The third issue is that of quality of life. In the past, the main concern has been over increasing the quantity of travel, the acquisition of a car and the notion of the freedom to use that car. As affluence increases, other factors related to quality and environmental responsibility become important, and values change. The technological revolution now taking place allows such a transition. From the viewpoint of transport planning, the imperative to predict the growth in demand and the overriding importance of economic factors in assessment now becomes less dominant. Methods need to be developed to measure quality of life, social impacts, and the environmental/ecological costs of transport. This change in societal priorities should mark a move away from the necessity to quantify everything and to ignore or devalue those factors which cannot be quantified.

Such a switch would result in the replacement of the 'Anglo Saxon' approach to analysis with its narrow conceptual framework and its allocation of a market role to transport with efficiency and productivity being the prime objectives. In this regime, intervention only takes place to correct market failures or to make adjustments for social reasons. Perhaps a more European approach to transport planning will be adopted which will assign transport the status of an intermediate activity that requires direct control to achieve wider social, industrial, regional and national objectives. To this list would be added quality of life and international objectives.

The optimistic scenario presented for the new century is that transport planning will undergo a renaissance, in a modified form as the requirements of strategic planning, the new realism, and the quality of life imperatives dictate. At last, the TPM appropriate for application to large cities in the 1960s and 1970s has been retired. Although mainly dormant in the 1980s, the systems approach was still influencing thinking, but events in politics, in the growth of demand, in new research agenda, and in public attitudes have all resulted in a fundamental change. Such radical shifts have occurred before, but never has such a consensus existed. It is this new thinking and agenda which are addressed in the second part of the book, but first there is a diversion into international experience to establish whether there are lessons to be learnt from their approaches to essentially very similar problems.

Note

1   ESTEEM – Estimation of Travel, Energy and Emissions Model. This is a GIS based model developed at the Bartlett School of Planning under two EPSRC grants and it is available for use by local authorities. It uses secondary data and runs with ArcView and NetEngine GIS software supplied by ESRI.

Chapter 7

Overseas experience

Introduction

The problems outlined in the first part of the book are not restricted to Britain. Urban transport planning throughout the developed world is now faced with a series of paradoxes and dilemmas which have been neatly summarized by Hall (1985).

1 From the mid-1960s onwards, urban transport planners almost everywhere forsook the ideal of individual motorized mobility for all. They substituted a combination of urban traffic restraint and the promotion of good public urban transport.
2 Their change of mind (or heart) was fortified in the 1970s, first by the limits of growth arguments contained in the Club of Rome report of 1972, second by the living manifestation of that argument in the great energy crisis of 1973–74, and third by the appreciation of ecological impacts throughout that decade and into the 1980s.
3 The first two of these influences are now much diminished in force. Soon after 1973–74, energy prices resumed the broad secular downward courses (in relation to general price levels) that they have manifested, with only brief perturbations, throughout the entire post-World War Two period. The ecological argument in contrast looms stronger than ever, though its precise significance for transport policy is not always clear and is certainly debatable.
4 Meanwhile, throughout the 1970s, there was a clear and very general tendency towards dispersal of population – and, in some cases, economic activity – from cores to peripheries in urban areas. There are wide variations in the extent and the time of this process as between one country and another, but it may be that these merely represent points along a common trend line. If the pattern continues, it can only strengthen the use of the car and weaken the position of public transport authorities.
5 In any event, the financial position of these public transport authorities has in general deteriorated because of the growth of car ownership and use. The resulting central and local governmental subsidies are now being challenged because of the pressure of the recession on public expenditure.

In a slightly different form, these paradoxes are still apparent, and the dilemma between the desire to take advantage of the new mobility brought about by car (and air) travel has not been fully reconciled with the environmental costs and the distributional consequences created by the car. There are the same debates over the role of the car (particularly in cities), the carrying capacity of the planet (particularly the levels of CO_2 emissions), the pricing of fuel, the questions of concentration and dispersal of cities, the financing of public transport investment, and subsidy for all forms of transport. The context is, however, very different from that in the 1970s and 1980s, as globalization is a reality with international competitiveness adding to local and regional competition. Society is moving from a manufacturing base to a service and knowledge based economy, whilst the role for technology is still tantalizingly unresolved. It is important to learn from experiences in other countries, hence there is a slight diversion with a comparison of thinking and approaches used in some other European countries and the United States.

Transport policy in the European Union

Policy to 1995

The EU provides a clear example of the choices that need to be made. In its Common Transport Policy (CTP) statement of 1992 (CEC, 1992*b*), one of the main themes was Trans-European Networks (TENs). Incompatibilities between national transport systems have been highlighted, including inadequate interconnections, missing links and bottlenecks, and obstacles to inter-operations. All of these resulted in inefficiencies, but the EU played only a limited policy role mainly through the Committee on Transport Infrastructure (set up in 1978). The EU now proposed to establish and develop a

> Trans-European transport network, within a framework of a system of open and competitive markets, through the promotion of interconnections and inter-operability of national networks and access thereto. It must take particular account of the need to link island, landlocked and peripheral regions with the central regions of the Community. (CEC, 1992b, para. 140)

The goal was to improve the integration of the Community transport system and not the improvement of the transport infrastructure in general. It was likely that much of the funding would continue to be allocated to the geographically isolated regions. On the crucial question of financing, the White Paper (CEC, 1992*b*) was pessimistic. The general level of investment on transport infrastructure in the EU had been stagnant at about 1% of GDP. The volume of investment required for the period 1990–2010 is nearly 1500 billion Euros, or 1.5% of GDP (CEC, 1992*b*, para.143). This level was far in excess of the resources available to the EU, even if its mandate would permit such intervention. Its role was limited to financing feasibility studies, loan guarantees, and interest rate subsidies. In addition, the EU had a major dilemma. On the one hand, it saw under-investment in transport infrastructure, but on the other hand it was also arguing for sustainable mobility and protection of the environment (Table 7.1). It could be argued that these two objectives were incompatible (Ross, 1998).

It should be acknowledged, however, that EU policy was more of a stimulating and initiating nature, while the final responsibility for implementation and enforcement still largely rested with the individual Member States of the EU. At the same time, the CTP had become increasingly important, due to several 'package deals' (Hey, 1996). First, as a result of the internal market objective, transport markets were liberalized and the Trans-European Networks became a pre-requisite for the functioning of the internal market. Second, from 1987 onwards more financial support (such as for the financing of transport infrastructure) was granted in order to compensate peripheral regions for the negative impacts of the internal market. Third, these funds became larger in 1992 in order to compensate these regions for economic disadvantages of the convergence criteria necessary for the single currency.

In summary, there were three main elements to the CTP in 1992:

+ *Efficiency*: subsidies should be reduced and market principles should be increased in the operation of the transport system and the assessment of new investments; in this way the transport system should contribute to economic efficiency of society and to an improvement of the competitive position of the economy.
+ *Regional development*: the transport system is a means to stimulate

Table 7.1 Objectives of the Common Transport Policy

- The continued reinforcement and proper functioning of the internal market facilitating the free movement of goods and persons throughout the Community.
- The transition from the elimination of the artificial regulatory obstacles towards the adoption of the right balance of policies favouring the development of coherent, integrated transport systems for the Community as a whole using the best available technology.
- The strengthening of economic and social cohesion by the contribution which the development of transport infrastructure can make to reducing disparities between the regions and linking island, land-locked and peripheral regions with the central regions of the Community.
- Measures to ensure that the development of transport systems contributes to a sustainable pattern of development by respecting the environment and, in particular, by contributing to the solution of major environmental problems such as the limitation of CO_2.
- Actions to promote safety.
- Measures in the social field.
- The development of appropriate relations with third countries, where necessary giving priority to those for which the transport of goods or persons is important for the Community as a whole.

---

economic development in more peripheral regions (especially CEC-countries and Southern Europe) and is used to stimulate the social cohesion within Europe.

♦ *Environment*: the transport system has to reduce its external (environmental) impacts, so that the system favours a sustainable (environmental) development.

Although most of the CTP had been focused on infrastructure, this transition phase (1992–1995) also brought about other important changes in regulations, particularly on safety and the environment (CEC, 1995). European regulation aimed at reducing air pollution by road vehicles by setting emission reduction targets per vehicle, by reducing traffic congestion, and by reducing mobility growth. The first had been relatively successfully applied over the recent past in Europe, which reduced emissions of several gases by up to 50% (OECD, 1993). However, the simultaneous rise in mobility has meant that the net energy consumption and emissions of CO_2 by transport have increased. The reduction of CO_2 emissions is seen as a major environmental challenge, both in terms of the original stabilization target agreed at Rio (to reduce CO_2 emissions to 1990 levels by 2000) and the subsequent target set at Kyoto (to reduce CO_2 emissions by 8% between 1990 and 2008–2012).

Policy since 1995

In 1995, the Commission launched its action plan for 1995–2000. As part of this new initiative, there has been a series of important debates opened up in the transport sector. Although the main aims of the CTP of 1992 have not changed fundamentally, there was a significant change in the focus of transport policy in the EU. The efficiency of the transport system was still central to much of the policy thinking, as this was seen as being essential to the competitiveness of Europe and to growth and employment. But a greater emphasis was being given to the social cohesion objectives, to safety (again), the environment, subsidiarity, and the accession countries. The three main policy themes were:

Improving efficiency and competitiveness of the transport system was concerned not only with new infrastructure and the completion of the TENs, but with four other main policy initiatives:

♦ liberalizing market access (particularly as it related to railways, air and ports);

♦ ensuring integrated transport systems across Europe (continuation of the TEN-Transport priority projects, but with public private partnerships for financing and operating these systems);

+ ensuring fair and efficient pricing within and between transport modes, in particular applying the principles of marginal social cost pricing;
+ enhancing the social dimension so that more balanced and sustainable development can be implemented across all the EU.

Improving quality in response to the needs of EU citizens means that priority was given to the following three areas of policy:

+ Safety is a permanent concern of the EU in all forms of transport, particularly in the air, maritime and roads sectors.
+ The development of sustainable forms of transport to limit the impact of transport activity on climate change. This work includes the development of accurate indicators of transport and the environment, and the strengthening of the environmental impact assessments of policy initiatives. Links are being made here with air transport noise and emissions, with waste reception in maritime transport, with the problem of heavy lorries in the roads sector, and with the emissions work of the Auto/Oil I and II programmes.
+ Protecting consumers and improving the quality of transport services through participation and representation of organizations in the development of the CTP. The two main sectors concerned here are in aviation and local public transport. In the latter role, a Citizens Network has been set up to establish best practice, including the integration and benchmarking of public transport services.

Improving external effectiveness covered the links with the accession countries and the globalization of the world economy. Agreements have been negotiated with some of the accession countries so that markets can become more open during the transition period to membership of the EU. The enlargement of the EU must be achieved with minimum disruption. The globalization issues concern trading and market conditions as they relate to external countries, particularly the United States, Switzerland and other countries.

As can be seen from the discussion above, since 1992 the CTP has evolved substantially to a much broader-based and more coherent approach (CEC, 1998*b*). The primary concerns of policy within the EU along the three original dimensions of competitiveness, cohesion, and the environment are still present. They formed the first two of the new priorities (efficiency and competitiveness, and improving quality), but the two new dimensions relating to the accession countries and the role of the EU in global markets have substantially enhanced the scope of the CTP by improving external effectiveness.

Secondly, the original concerns were primarily with the network and the means to provide a European infrastructure to link all the EU countries together, and to link with the countries of Eastern Europe (CEEC) and the Soviet Union (CIS). This has also changed with a new emphasis on bringing down the barriers to free trade (and using pricing tools more effectively), making the systems compatible (interoperability), getting the best out of the different modes of transport (intermodality), making good use of the network (interconnectivity), promoting best practice in organizational structures (including logistics and technology), and in ensuring the responsible use of resources in transport. Strong links have now been drawn between the transport policy perspective and the new European Spatial Development Perspective (ESDP), as the combination of these two policy areas is necessary to achieve sustainable mobility and a balanced territorial development.

However, there may still be inconsistencies in EU transport policy, particularly as it relates to the environment and the achievement of the challenging Kyoto targets for CO_2 reduction. As it states in the Communication on EU CTP (CEC, 1998b)

> it will be necessary to assess more globally to what extent existing policy measures will bring the transport sector in line with environmental objectives and what further well-focussed and complementary measures may be needed. Particular attention will need to be given to measures designed to reduce the dependence of economic growth on increases in transport activity and any such increases on energy consumption, as well as the development of less environmentally damaging energy alternatives for transport. (para 46)

These are the new challenges of the Common Transport Policy for the next 5 to 10 years. It is a dynamic instrument for delivering an integrated transport system that both accommodates the needs of the planet and those of all its inhabitants. The policy discussion paper (CEC, 1998a) concludes that much progress has been achieved, but to sustain economic progress, social structures and a clean environment, significant further agreement at EU level is required. In the recent EU Draft Transport Policy document (CEC, 2001), the focus switches to decoupling transport growth from economic growth. The aim of a sustainable transport system is to enhance economic growth but with less transport, so that the balance between transport modes can be redressed and pressure on the environment can be relieved with reductions in the levels of congestion.

In Britain, there has been a move towards greater efficiency through the use of the market mechanism and a reduction in levels of public support for transport. In other countries (for example, France, Germany and the Netherlands), transport is seen as part of a wider social policy where allocative efficiency is more important than cost efficiency. These different traditions are highlighted here with the European traditions being compared to those developed in Britain and the United States. The traditions of transport planning and urban planning in Britain and the US are directly comparable, even though the policy context both politically and economically may have been different (Chapter 2). In the next sections, some of the main elements of transport planning in European countries and the United States are presented, both for comparison purposes and to ascertain whether there are common lessons to be learnt.

Transport planning in Germany

The main priorities in Germany have been the reconstruction of the infrastructure after the reunification of the country in 1991. Before then, in the early 1960s there had been an increase in car ownership, but it was only in 1965 that Germany had about the same number of cars as Britain (8.5 million). The tradition in transport planning was to collect all the relevant data through traffic counts, origin and destination surveys and traffic densities so that forecasts could be made. The techniques and practice developed in the United States were used as German traffic engineers had limited experience themselves (Hass-Klau, 1990). Even though public transport was still the main mode of transport in large cities there were arguments about whether the trams should be replaced as they interfered with the free movement of the car. The free market ethos extended to road transport and the journal *Verkehr und Technik* wrote in 1953 'constructing roads is not only an issue for the car users; it is an issue for the whole nation'.

The standard book used at this time was Leibbrand's *Verkehr und Städtebau* (1964). It favoured the separation of traffic as this allowed faster speeds and greater safety, and argued that the car was the salvation of the city. Free access to the city centre would allow growth in the urban economy and the maintenance of prosperity. It seems that the Germans were more car oriented in their thinking and this was reflected in their road building programme. However, even at that time concerns were arising over the implications of the continued programme of road building. The second major report *Die Kommunalen Verkehrsprobleme in der Bundesrepublik Deutschland* (Hollatz and Tamms, 1965) brought together the leading German researchers to review the problems of traffic and their conclusions were still clearly in favour of the car:

> it cannot be considered to restrain private motor transport: a rational design for urban traffic was needed. A sensible division of different transport modes was necessary, which implies traffic restraint in particular locations.

The report recognized that urban structure had been influenced by the car and proposed a system of tangential and orbital routes around cities. Traffic had to be controlled in city centres, principally through parking restrictions, and public transport should have a separate right of way.

Strangely, there seems to have been little interaction between the Germans and the British as the Buchanan Report *Traffic in Towns* was being prepared at the same time. Similarly, in the United States there was a major shift in thinking with the establishment of the 3C Planning Process (1963) and the Urban Mass Transportation Act (1964) which allowed the use of federal funds for capital projects in public transport for the first time. In 1969, a federal law in Germany established a 15 year plan for the completion of the federal highway system.

Prior to 1967 public transport did not receive any funds from the federal government, and by 1965 it had reached a low point in terms of efficiency and the numbers of passengers carried (Hall and Hass-Klau, 1985). Even at this early stage, a change in philosophy was detectable as urban planners were active in the pedestrianization of city centres – over 60 cities had introduced schemes (Monheim, 1980), including Bremen with its system of four traffic cells where movement by car could only take place within each cell and not between cells. The Federal Act (1967) increased petrol tax (*Mineralölsteuer*) by 3 pfennig a litre with 40% of the revenues being used to finance public transport investment. This proportion was later increased to 50%, but by 1989 it was back to 40%. In Germany, the road lobby has always been particularly powerful and apart from restricting the amount of money transferred to public transport from the petrol tax, it also managed to get a dense network of interurban motorways approved so that 85% of the population would be within 10 km of the system.

The end of the 1960s was also marked by student revolts which (as in France) changed politicians' attitudes towards urban renewal and public transport. It was also the first time that the Social Democrats had come to power (1969), with the expectation that liveable cities and a more socially aware society would be created. The dominance of the car in that society was beginning to be questioned by Dollinger (1972) and Dahl (1972). Again, as with other countries it seems that there was a window of optimism prior to the global events of 1973, which marked an end, at least temporarily, to the growth in incomes and prosperity.

As in Britain and the United States, Germany made use of the aggregate models of land use and transport, establishing relationships between people, motorization and mobility, predicting changes based on population trends and developing a five-stage sequential demand model. The desirability of a trip or attractivity of the place was established from the population and social status of the location, and this was followed by the conventional trip generation, trip distribution, modal split and assignment stages (Mäcke, 1964). The transport analysis was then followed by a functional analysis of the transport system and the city as a whole. Regression techniques and gravity models seemed to be the most commonly used techniques, with cost benefit analysis being used to evaluate the alternatives.

Reaction against the use of aggregate models was triggered by Kutter (1972) who questioned the assumption that traffic forecasts could be based solely on objective data from the populations of areas. He argued that travel behaviour must be based on surveys of behaviour and attitudes. The assumption that people wish to travel from places of small 'mass' to places of large 'mass' is incorrect (that is, the notion of gravity does not apply to people's movements) as it postulates the independence of spatial characteristics and human motivations (Paradeise, 1980). With the use of disaggregate models the individual could be re-established as the prime unit of analysis which would allow analysis of constraints, transport being viewed as a derived demand. Such an approach would allow greater transparency and increase the predictive value of models by incorporating structural changes in population. Three types of disaggregate demand models have been developed:

- Econometric models which do not appear to have been developed in Germany. These have mainly been used in the Netherlands and the United States.
- Models based on the identification of homogeneous behaviour of groups of people. These have been extensively used in Germany.
- Situational approach which extends the modelling framework to include decision and learning theory. Again, Germany has been instrumental in the development of these stochastic methods which attempt an understanding of motivations, and the identification and measurement of preferences.

The professional involvement of the traffic engineer has always been a powerful influence on the development and use of techniques, and that professionalism was dominant in the 1960s and 1970s. Even when new methods were being developed, it was still the engineer who was using them. And when the social scientist became more involved in transport research, there were still institutional barriers to the acceptance of new methods as decision-makers came from the same traditional engineering background.

In 1973, the *Bundesverkehrswegeplan* (the Federal Transport Plan) provided a framework for all transport, including public transport. The catalyst for action had come from Munich where the first rapid rail system had opened (1972), and the extensive pedestrianization of the city centre together with new housing and development on the periphery received worldwide attention. Such schemes had been facilitated by the *Städtebauförderungsgesetz* (Urban Development Act, 1971) which promoted the renewal of inner-city areas with extensive federal support and gave active encouragement to stronger citizen participation in these decisions. By 1975, expenditure on public transport was about five times that in 1967, but the Federal Planning Reports at that time were still presenting the same argument as those used in the 1960s. The major reason for inner-city decline and the movement of people out of the city centre was the damaging effect of the car, which would according to the forecasts increase in the future. The conflict between transport and urban structure would sharpen (Hall and Hass-Klau, 1985). Two measures to counter this decline have been the promotion of public transport in city centres and the concentration of through traffic onto major roads to allow traffic free areas in city centres and residential areas. Yet the federal ministry did not allocate investment to support the public transport measures.

As noted earlier, pressure groups in Germany have always had a powerful influence on policy. In the 1960s and 1970s the road lobby maintained the heavy motorway investment programme, but it was the rise of the Greens and the environmental movements in the late 1970s which also influenced policy. In 1987 the Greens were the third most important party in the federal Parliament.

Some 7000 km of planned motorways were cancelled because of their negative effects on 'nature and landscape', and the priority for road construction switched to bypasses for small and medium sized cities. The importance of public transport has been extended from its promotion in urban areas to its retention in rural areas. Because of financial and budget restrictions, it now became important to evaluate alternative investments more systematically, principally through cost benefit analysis. For many public transport schemes it was also necessary to carry out a wide social analysis of benefits, particularly in rural areas and where there were wider development benefits. There was an undertaking to subsidize public transport for its social function. The subsidy to local passenger transport increased from 4 billion DM (1970) to 11.8 billion DM (1980), with the federal contribution falling from 68% to 58%.

The *Länder* play a key role in transport planning as funds for local road construction and local public transport support are distributed through them. Under the *Gemeindeverkehrsfinanzierungsgesetz* (Community Transport Financing Act, 1975) projects can be funded by up to 60% from the federal government. Over the period 1967–1981 some 23 billion DM were spent on S-Bahn, U-Bahn and Stadtbahn schemes in Germany, with the federal government funding 56%. The remaining 44% was split evenly between the *Länder* and local authorities (Girnau, 1983).

The concern over increasing levels of public transport subsidy has resulted in organizational changes and the creation of *Verkehrsverbunde* (Transport Associations) in the major German cities. These partnerships between transport operators have allowed the pooling of revenues and resources, the effective coordination of public transport services, the common marketing of services, and the integration of fares structures. Although the ultimate control remains with the separate operators and the public authorities to which they are responsible, there is considerable cooperation between operators established as a series of direct commercial agreements. All local and national operators are included and the partnership contract specifies the duties of the different partners including their rights and obligations. The revenue is pooled and distributed according to patronage and the costs of meeting the *Verkehrsverbunde* specifications. In Hamburg the HVV (the Hamburg Transport Association) replaced eight agencies responsible for the provision of transport (1965), and similar *Verkehrsverbunde* were subsequently established in Munich (1972), Frankfurt (1975), Stuttgart (1978) and the Rhine-Ruhr (1980). Such an organizational change allowed greater quality of service to be accompanied by increased efficiency, with cost recovery levels from fares being improved to 65%. A greater weight was placed on coordination of services than in deregulated Britain, yet it seems that efficiency was also improved and patronage levels maintained despite the size and complexity of the *Verkehrsverbunde*.

Since 1980, comprehensive transport plans have been produced for cities, for states and for the federal state. These plans have formed the basis of the coordinated transport investment programmes. The aim of the plans is long-term forecasts of demand with the objective of matching up supply of infrastructure for road traffic and public transport. The methods used have not been general land-use transport studies but more focused investigations, such as the effects of policies to switch demand from car to public transport or to non-motorized transport. Analysis considered information and marketing strategies as well as infrastructure related policies. Efforts have also been directed at developing multi-criteria evaluation methods to assess transport investments, including environmental pollution, land-use, traffic and financing aspects (Wermuth *et al.*, 1990).

Germany has also pioneered the concept of *Verkehrsberuhigung* (traffic calming) in city centres and residential areas. By the mid-1980s, it seems that most German towns with over 50,000 inhabitants had a pedestrianized area (Hass-Klau, 1990). It was realized that the car and the city were incompatible, and to maintain vitality of the city centre it was necessary to create 'liveable' streets (*Wohnstrassen*). Close parallels can be drawn with the Dutch *Woonerven* (see section on the Netherlands).

The planners had been instrumental in widening the debate on urban renewal from the concern over poor housing conditions and a lack of opportunity to consider the urban environment and the quality of life. The initiative had come from the Federal Ministry of Regional Planning, Housing and Urban Development (*Bundersministerium für Raumordnung, Bauwesen und Städtebau*) and was in effect a covert criticism of the Federal Ministry of Transport (*Bundersministerium für Verkehr*) which was still advocating maximum accessibility for cars. Much of the research has been carried out by urban planners, geographers and architects, not the traffic engineers, and the case was based on safety as well as environmental arguments.

Since the uneasy early alliance, traffic calming has become a broad based transport and planning tool which integrates all forms of transport with urban development. The 30 km/h speed limit in residential areas was combined with good design to ensure correct priorities were allocated to people and vehicles. More fundamentally, calming may have altered public attitudes to the car. Its dominance and attractiveness is not being questioned, but people and planners are no longer unaware of its limitations and the basic incompatibility between pedestrians and cars.

Germany may be moving towards a new transport policy (Zopel, 1991; Monheim, 1997*a*). Traditional transport policy and planning aims to avoid or remove bottlenecks by providing new roads and parking facilities. Provision for the car encourages greater car use and aggravates the problem it is trying to solve. What is needed is an environmental transport policy with public transport being provided as an attractive and pleasant alternative. Transport wastage is leading to a traffic and environmental crisis. A better quality of life and a better environment can only be achieved through an acceptable level of mobility and a high degree of efficiency. It suggests that Germany may be moving beyond the car, at least in the city.

Comment

The Germans seem to have been more active than the British in building new roads around and between cities, in substantial investment and subsidy in public transport, and by calming traffic in city centres and residential areas. Rigorous change has been implemented simultaneously on these three fronts, whilst in Britain and in France a more cautious approach has been adopted. The federal structure allows central government to set the framework for planning and transport, and to control strategic policy on public transport. The decentralized *Länder* have responsibility both for land-use controls and for public transport. In France decentralization of power has also given the regions greater autonomy, but in Britain the reverse has occurred with increased centralization and control over local authorities, through limitations on the ability to raise revenue locally and through cash limits imposed by central government on local expenditure. In Germany, the concept of subsidiarity is much more commonly used than in Britain. Decisions are made at the local or state levels, and each territorial unit is financially responsible for its fixed transport assets, including public transport. But the federal government still has a very strong influence in local investment decisions through project-linked funds. The 'golden rein' still operates in local decision-making.

Changes in approach have been brought about by powerful pressure groups. These include the transport enterprises, especially publicly owned firms, syndicates of trucking firms, trade unions, and automobile clubs. There are twelve million members of the automobile club and some 50 million drivers in Germany. Conversely, the rise of the Green party and the environmental movements has also been influential in developing alternative strategies for urban areas, including traffic calming and heavy public transport investment. These environmental groups have a secure political base and a broad based programme from which to operate. There are more than 2000 such action groups in Germany (Apel and Pharoah, 1995). In Britain, pressure groups tend to be more restricted in their actions, concentrating on particular issues. Their political base is weak and the electoral system does not allow them full representation in politics at central or local levels.

The strong engineering tradition in Germany in transport analysis is still apparent, and the conceptual framework established in the 1960s is dominant. Rather than involve social scientists in analysis (as in Britain) or to have a different tradition of social analysis (as in France), the German engineers have themselves diversified to embrace disaggregate methods and the situational approach. Social scientists have tended to concentrate on small-scale empirical surveys which are time consuming, expensive and difficult to make generalizations from. There is a need for a coordinated series of investigations where more general data are supplemented by a series of in-depth studies which would allow comparison and a common reference base to be set up – KONTIV is one example of such an approach. These data also indicate that although distances travelled in Germany have increased, the overall travel time of people has remained relatively constant (Brög, 1992). In addition these approaches have emphasized the public awareness issue with mobility patterns being published for many cities, together with summary comparisons. This information has been used to raise awareness among the population and to influence decision-makers for change. Transport planning has become a highly politicized process rather than a technical one, and this is consistent with changes in other European countries. The technical role of analysis has to be combined with the means to find the necessary consent among all interested parties. Many cities now have traffic forums (for example, Heidelberg, Münster, Salzburg and Tübingen) where bargaining between the different groups takes place under the control of a mediator (Monheim, 1997b).

Transport planning in France

Transport planning in France falls into two distinct phases, the first of which was concerned with road investment to catch up on construction after a long period of stagnation under rail-oriented state planning in the 1950s. The second, started in 1973, marked the rehabilitation of public transport. National planning in France had traditionally been top down with systematic data collection, forecasting and monitoring with the national institute (*INSEE: Institut National de la Statistique et des Etudes Economiques*) being the focal point of this activity (Benwell, 1980). There was considerable political power in the car industry and in all the National Plans produced in the 1960s and 1970s increased car output was anticipated. The Fifth Plan (1966–1970) demonstrated a 'Buchanan like concern with the need to improve highways and to adapt the town to the car'. The thinking was that the car was the only viable means of travel and huge urban investment plans were approved with the construction of inner ring roads.

It was at the end of the 1960s that financial limitations were imposed and other factors such as the quality of life and the indispensability of public transport became more prominent. The first bus priority schemes had been introduced in Paris and Marseilles in 1964, and political pressure was imposed by public transport operators for increased government investment. The *Loi d'Orientation Foncière* (Land Guidelines Act, 1967) permitted the grouping of adjacent local authorities for planning purposes so that informal strategic plans could be drawn up. Perhaps the most important single factor which triggered many changes in attitude and policy was the student riots of 1968.

In 1970 the *Colloque de Tours* called for integrated multi-modal transport and a rebalance of investment towards urban public transport. This meeting was the first political initiative of the public transport lobby since 1945. The professional focus at the *Ponts et Chaussées*, the main training centre for French traffic engineers, was still oriented towards highway construction and traffic planning, but with the enlarging of the profession and political pressure decisions were reversed. A few critical urban motorway proposals were abandoned such as the left bank of the Seine expressway through Paris, and large-scale public transport investment took place in Paris with the first *Réseau Express Régionale* – RER line, the complete refurbishment of the Métro and the bus fleet. In addition the decision was made to construct other RER lines and métros and light rail networks elsewhere. The Marseille métro was opened in 1977 and the Lyon métro in 1978. Public transport professionals were able to protect their programmes through contracts between agencies (*Régie Autonome des Transports Parisiens* – RATP and *Société Nationale des Chemins de Fer* – SNCF) and devolved planning agencies replaced direct government control and supervision.

The 1970s was the decade of public transport with the Sixth Plan (1971–1975) and the Seventh Plan (1976–1980) both marking heavy investment in the public transport network and the balancing of financial objectives by social ones. By the end of the decade energy savings were also considered important as France imported over 75% of her energy. Local transport planning was renewed in 1973 (*Dossier de Transport*) with a 10 year strategy statement and a 5 year programme with a priced tactical programme, and consisted of a set of long-range plans heavily oriented towards road and motorway construction. Each town with a population over 80,000 was to prepare a plan. Local authorities took over ownership of the public transport network and a local transport tax paid by employers, *Le Versement Transport*, was introduced in 1971 in Paris. In 1973 it was extended to cities over 300,000 and in 1974 to towns over 100,000 (Table 7.2). By 1988 *Le Versement* was funding 30% of the costs of public transport (users 45% and local government 25%). The *Carte Orange* was introduced in Paris and there was greater user representation in decisions. 'Urban transport policies in the 1970s can be analysed as local takeovers of national procedures' (Lassave and Offner, 1989). This municipal takeover of urban transport resulted in a significant improvement in the quality of public transport, the protection of town centres against the car, and an attempt at the evaluation of non-motorized transport and the needs of particular social groups.

The separate plans were produced for each town of more than 10,000 inhabitants. A long-term land-use plan (*Schéma Directeur d'Aménagement et d'Urbanisme* – SDAU), which had been in use since the late 1960s, looked 30–40 years ahead and identified broad land use zonings. By 1980, some 370 SDAU had been started, covering 26% of the area and 70% of the population. As with the comparable structure plans in Britain, progress was very slow even though they were more modest in scale. A medium-term land-use plan (*Plan d'Occupation du Sol* – POS) zoned all land including conservation areas, urban development land, and locations with special controls. By 1980 some 11,400 POS had been designated (Simpson, 1987). Transport studies were complementary to these zoning plans and can be seen at five levels (Ferreira, 1981).

- Long-term strategic studies interacting with the SDAU and financed by central government. The main purpose of these studies is the identification of the major corridors for transport in cities and towns (over 20,000).
- Medium-term studies (20–25 years) to determine the exact land requirements for a specific facility. These studies must take the medium-term land-use plans (POS) into account.
- Investment programming studies. These are 5 year plans containing a detailed priority list of transport improvements. Such studies are carried out at the city level and updated annually (*Dossier de Transport*).

- ◆ Detailed costing studies of specific projects.
- ◆ Detailed design and documentation for specific projects.

Since the global and national economic transformations of the 1970s, the end of long-term simple mechanistic forecasting methods was becoming possible. The social nature of communications needs was now recognized as the basis of physical movement. However, traditional methods are difficult to replace completely and although it was now possible to take a broader social-based approach to analysis, there were still aspects of these methods in evidence. The change was not clear cut. The 1980s marked this change with a decentralization of decision-making. France's modernization of local government came very late (1982–85) and the 36,000 communes have still been retained. Local authorities already had powers for public transport, traffic, parking and town planning, and now land use and traffic management were added.

The *Loi d'Orientation des Transports Interieurs* (LOTI, 1982) was directed towards 'the right to transport for all', as all users should now have the right to travel and the freedom of choice of mode with reasonable access and at a reasonable cost. Priority for public transport was to be balanced by complementarity with other modes and by fair competition. Urban areas were encouraged to prepare urban travel plans (*Plan de Déplacements Urbains –* PDU) with a statement of policy to cover traffic systems and priorities for all modes of transport. In theory, part of the thinking here was to reverse the trend towards urban sprawl, to give priority to public transport to increase its efficiency, and to reduce the dependence on the car. The PDU also introduced a public inquiry procedure to open up the debate between the population and the professionals and politicians, but also to get their agreement and support for proposed policies.

Within LOTI (1982) there was also a requirement for an infrastructure master plan to cover an evaluation of alternatives including social and economic factors, and this document should be made public before the adoption of a project. The *Schémes Directeurs d'Infrastructures* are prepared by the state in consultation with the regions and by the municipalities (Simpson, 1987).

The *Versement Transport* was also extended to all towns with populations over 30,000 (1982). As with the larger settlements, the implementation of the tax was optional, but most local authorities took the opportunity to introduce it at the maximum permissible levels. All employers with more than nine wage earners must pay the tax, and the levels are calculated on the basis of salary ceilings. In 1988 the level was 9950 FF/month and the application of fixed percentage rates (Table 7.2). In Paris most of the revenue has been used to pay for tariff reductions given to employees (*Carte Orange*) with little contribution to depreciations and nothing to finance new infrastructure. Receipts in Paris have started to decline in real terms as decentralization takes place. If additional revenue is sought, the only option would be to raise the fixed rate again. Elsewhere, of the 125 possible locations, 104 had introduced the *Versement Transport* (1986) with some 60% of revenues being allocated to revenue support and the remainder being used for capital investment and the repayment of interest on loans.

Table 7.2 History of the Versement Transport

1971	Inner Paris	1.7%
1973	Cities over 300,000	1–1.5%
1974	Cities over 100,000	1–1.5%
1975	All Paris Transport Region	1.2%
1982	Towns over 30,000	0.5%

Note
Percentages give the amount that can be raised by local authorities on the salary ceilings

The *Versement Transport* allows some continuity to public transport funding as the operators know the approximate levels which will be raised each year. The salary ceiling and the level of the tax may be changed, but that continuity allows longer-term decisions to be made. In the United States and Germany the tax on petrol with an allocation to public transport is an alternative way to fund loss-making bus and rail services, and to encourage investment. In Britain there is no such continuity and public transport operators are dependent on fare revenues and subsidies received from local authorities for contracted services. Privatization of bus services has meant that operators can now raise capital on the commercial markets, and this has resulted in substantial new investment in buses, at least from the larger operators. Much of the explanation for the poor quality of the British transport infrastructure, particularly in public transport, can be explained by the failure of successive governments to establish a stable financial basis for long-term investment, either through an employment tax or through a petrol tax, or even through facilities to borrow capital at attractive interest rates as is permitted in France. To some extent this problem has been addressed in the Transport 2010 document (DETR, 2000*b*), but its impact has yet to be seen, even though public services and investment are at the top of the political agenda in the United Kingdom.

There has also been a switch from the focus on urban to interurban applications of formal evaluation and modelling. Urban evaluation has become very pragmatic for small projects with a strong qualitative orientation, and it can be seen (as in Britain) very much as a bidding process for resources. Large projects such as the *Train à Grande Vitesse* (TGV), the Channel Tunnel and the RER lines require more classical and formal models.

The TGV Sud-Est line between Paris and Lyon (427 km) was opened in

1981–83 with a further extension to Grenoble in 1985 to carry passengers at speeds of 270 km/h along new specially constructed track. The cost was FF 8.5 billion (1985) with a further FF 5.7 billion (1985) for rolling stock. The journey time between Paris and Lyon was cut by 1 hour 50 minutes to 2 hours, and traffic increased by 45% to 14.3 million (1983–84) and a further 10% to 15.8 million (1985). Formal economic evaluation calculated an economic rate of return of 15% and a social utility rate of 30%. The social utility broadened the evaluation to include the public interest and gave a value about twice that of the financial viability. These factors included safety and the environment, as well as structural changes which may be induced in regional planning and development. The capital was raised through borrowing by SNCF after the project had been approved centrally. There was also an inquiry prior to the formal approval by the state so that individuals could make representations on the basis of a report presenting the project and including an impact study. SNCF has excellent credit standings on the international market because of its government guarantees (Gerardin, 1989a, b).

By 1984, the TGV Sud-Est had generated FF 3 billion in revenue of which 35% covered operating costs and there was a return of FF 400 million after debt servicing and depreciation costs. The levels of traffic exceeded predictions with the new high-speed railway attracting 33% of its patronage from air, 18% from road, and 49% from induced traffic (including rail) (Bonnafous, 1987). Apart from the economic and financial appraisals, the TGV was also subject to environmental and regional impact analyses. For the services to be profitable the links had to be from Paris and this has meant that links between regional centres have not been improved. Paris has become the national hub. A series of surveys were carried out along the Paris–Lyon corridor and these indicated growth in tourism generated by the TGV as well as growth in high level service industries (Bonnafous, 1987). However, the TGV has had little effect on the places along the route (at least initially), only a powerful one at both ends. The Paris–Lyon link broke the psychological barrier as the return journey could be made in a day. The development impacts have been selective and seem to have reinforced existing trends. The TGV is a communications device which has changed the face of transport, but it alone does not change business decisions or the distribution of activities.

Subsequent decisions to build a TGV network in France were predicated on the success of the TGV Sud-Est. The French government adopted (1989) a master plan for high-speed rail links (Table 7.3) and other links are being proposed with lower internal rates of return and social utility. With the extension of the TGV line to the airport at Satolas (NE Lyon) for the 1992 Winter Olympics (Albertville), a choice from four routes was made. These options were compared by three experts appointed by the Ministry of Transport and they recommended the rural route. Counts were made of buildings between 40 and 200 metres from the axis and every effort was made to place the new railway next to the existing Lyon–Geneva motorway to minimize the environmental costs. The final decision of the Minister can only be challenged by instituting action in administrative courts and this may result in more protection without altering the layout, for example by the more extensive use of tunnels. Once the decision has been made the time between decision to proceed and opening of the project is short (4–5 years).

Table 7.3 The development of the TGV network in France

Title of TGV	Route	Length (km)	Opened	Cost (1985)	IRR (%)	SU (%)
Sud-Est	Paris – Lyon	427	1981–83	13.2	15	30
Atlantique	Paris – Bordeaux	580	1989–90	16.5	12	20
Est	Paris – East France and Germany	460		23.0	3.5	9
Nord	Paris – Brussels, London, Cologne, Amsterdam	300	1993	28.5 (1987)	7	11
Méditerranée	Lyon – Marseille	240	2001	26.0 (2001)		

Notes
IRR is the Internal Rate of Return
SU is the Social Utility
Costs are in billions of French francs

The latest addition to the network is the line from Lyons to Marseille opened in May 2001, and it has 'shrunk' the country as it is now possible to travel from Paris to the Mediterranean in three hours. As in Lyon Part-Dieu, it is expected that investment will now take place in Marseille to reduce the historically high levels of unemployment and to promote regional development. One of the main aims of the TGV has been to 'boost the local economy' (Bertolini, 1998), but others suggest that the impact may only be redistributional as appropriate local economic conditions are also important (Troin, 1995). On the Sud-Est line, development has only taken place at Part-Dieu in Lyon, not at Mâcon or Le Creusot, whilst on the Atlantique substantial growth has been initiated at Le Mans, Nantes and Vendôme. In most cases a 20% premium has resulted from a combination of access to the TGV, land availability, and buoyant local economic conditions (Banister and Berechman, 2000).

In France there is a broader based approach to evaluation of the TGV network which balances the economic and financial costs and benefits against regional development and environmental impacts. Financing is a complex issue as regional and European contributions have to be balanced against the part that SNCF could (or should) pay for. In addition, the assessment must ensure that new links do not undermine the viability of the network as a whole as the potential for traffic and revenue generation is considerable. The public inquiry stage is limited to debates on actual routes and the use of expert advisors is standard practice in both road and rail investment decisions.

The difference between the British and French system is that in Britain proposals are produced for local reaction whilst in France proposals are agreed and then discussed locally. It seems that France has developed a set of procedures and the organizational and financial framework necessary for a new generation of rail investment to produce a Europe-wide high-speed rail network. But the public sector model used in France does have its problems. There are two intercity rail operators – *Réseau Ferré de France* (RFF), which runs the infrastructure and decides on investment and maintenance and *Société Nationale des Chemins des Fer* (SNCF) , which is the public monopoly operator of services. Even though both operators have the duty to break even, they are in debt and rely heavily upon borrowing to meet their repayments. RFF charges SNCF a rate per train kilometre to run its services, and this amount varies according to type of train. The intention is to set the rate at a level that encourages increased efficiency on the part of the train operator, but at the same time allows the track provider sufficient resources for investment and to reduce its debt. This separation of track and operations took place in September 1997, but by 2000 RFF had made a loss of 3.5 billion Euros, and this was in addition to its long-term debt of 23 billion Euros. In 2000, RFF borrowed 18 billion Euros on the international capital markets to refinance its debt, but this means that it still has to pay 2.4 billion Euros a year in interest charges and 2.6 billion Euros to SNCF for managing the rail system. The French government has asked RFF to reduce its borrowings by 50% over the next 10 years, and there is also pressure from the EU to allow new operators to compete with SNCF in running train services (Quinet, 2000).

Comment

Transport planning analysis in France has had less emphasis on formal mathematical models and a much greater focus on simple empirical methods. Some modelling is required for the National Plans (carried out by INSEE), for the SDAU Plans, the more detailed *Dossier de Transport* and for all large projects. The urban division of SETRA (*Service D'Etudes Techniques des Routes et Autoroutes*) has compiled a set of manuals for strategic transport planning where accessibility has to be improved without any loss of quality of life. However, as early as 1973, questions were raised about massive data collection exercises and there were doubts in some circles about the relevance of large-scale models to the development of transport policy (Dupuy, 1978). France has never fully believed in the systems approach to transport planning, and has placed a greater emphasis on demand oriented analysis.

Trip generation and attraction rates were usually empirically derived and used in a gravity model to determine trip distributions. For example, the gravity model was used for the evaluation of the TGV Sud-Est railway, based on the generalized costs of travel between Paris and Lyon. Modal split was carried out on the basis of past trends, the options being tested with empirically derived utilization curves. Studies of central area parking demand and supply were carried out and the highway network examined to see whether capacity was exceeded. Cost estimates were made of the changes required to maintain the level of environmental quality and accessibility. There was feedback here to the modal split analysis. Assignment was carried out by the simple procedure of allocation of trips between each zone to the network by mode. There was a considerable amount of judgement used by the analyst in allocating traffic between two similar routes. For public transport, the network was defined to be consistent with the SDAU (*Schéma Directeur d'Aménagement et d'Urbanisme*) and the requirement to maintain good central area accessibility. Options were developed on the basis of demand using average figures per head of population, on accessibility as measured by trip times from home to work, and on urban structure.

In all these procedures there is a close link between land use and transport. Transport plans, both long term and medium term, form an integral part of urban planning and are incorporated into the land-use plans (*SDAU* and *Plan d'Occupation du Sol*). In preparing the SDAU, estimates are made of the amount and distribution of economic activity as well as the distribution of population. Other features such as public buildings, parks and recreational areas are identified. The transport studies estimate future peak hour traffic volumes and central area parking requirements. Future infrastructure needs are then estimated under a range of alternatives from heavy transit to private transport. A dialogue between politicians, planners, engineers and economists then follows to arrive at the best possible compromise solution.

This summary, taken from Ferreira (1981), leads to the conclusion that transport planning links with urban planning are stronger in France than in Britain, but that this consistency is at the expense of internal coherence of the transport planning process itself. Rather than having two separate analysis processes with little interface between them, the French have followed a strong planning philosophy with transport analysis being brought in to help identify problems and to test possible solutions. This is one attempt at a better integration of land-use and transport analysis. The plan is undertaken first (the *schéma de principe*) and the detailed analysis and evaluation is then embedded within it to establish the details of the scheme itself.

In addition, the *Plan de Déplacements Urbains* (PDU) mentioned earlier determines the organization of transport inside cities, in particular decisions on infrastructure investments and management, and the means to fund them. The PDU is a medium-term scheme, related to the longer term master plans, and it also identifies goals for transport, pollution and land use within the city, so that scenarios can be developed as to future policy directions. The scenarios are prioritized through indicators relating to travel costs, numbers of journeys, journey times, safety, total air pollution, noise, and land-use changes (Quinet, 2000). The actual evaluation process (mainly cost benefit analysis) is used to help identify priorities, but it has to be balanced against political pressures and the interests of different stakeholders.

This fundamental difference in approach can be explained by a series of related factors. Substantial transport research in France has been based in non-technical research traditions of sociology. As Benwell (1980) comments the French have lacked 'the impetus of the background questioning and assumptions which have led many US and UK researchers from mathematical and engineering backgrounds into econometric and attitudinal studies. Many French researchers have extended their interests into a transport context from a background in other branches of the social sciences'. The modelling problem-solving orientation of much British and American research is re-placed by a more consciously theoretical conceptual orientation of linkages between transport and land-use dynamics.

This different theoretical background allows many different perspectives on the behaviour of individuals, the role of cultural factors, and the organiza-tion of society to be assessed in the travel decision. At the macro level it also allows explanations of major policy changes. For example, it is generally recognized that fundamental changes took place in the early 1970s on investment policies for the Paris Métro as a result of large-scale user protests and pressures exerted by the Left. It also allows analysis of perceived irrationality of some public transport policies which were adopted. The RER A line was built to make the La Défense business complex viable. The needs of manual workers who had to move to the outer suburbs as a result of the increasingly homogeneous gentrification of inner Paris were not considered (Dunleavy and Duncan, 1989).

The focus was often on small-scale intensive studies once it was realized that sophisticated forecasting methods were not needed. Studies usually only took about six months to complete and often there were non-experts on all technical committees which meant that the techniques had to be simple and transparent. The models developed were not disaggregate or behavioural, yet the results seemed to be no less accurate than the more sophisticated approaches. A study which compared observed with modelled values from 17 home interview surveys in France found that errors ranged from 7–15% of the various trip generation categories, and that for trip distribution the errors were 25%. These figures compare favourably with those found in Britain (see Chapters 2, 5 and 6).

Transport planning in the Netherlands

It is in the Netherlands that transport planning seems to have maintained its position as central to environmental and national planning policy. Rather than moving away from planning as an approach to problem solving, the Dutch have produced (1989) a second National Transport Structure Plan (*Structuurschema Verkeer en Vervoer II– SVV-II*) to cover both passenger and freight transport to 2010. A series of scenario objectives has been set so that regular monitoring can take place and modifications can be made in the more detailed implementation document, the Infrastructure and Transport Programme. Specific policies fall into four categories, each of which is of equal importance:

Category 1: Improving accessibility – the provision of a high grade network of national and international roads, railways and waterways serving the goal of *Nederland Distributieland*. As national income rises, the transport industry's share can be held at not less than 7%. Transport proposals must be linked in with patterns of development proposed in the Fourth Report on Physical Planning.

Category 2: Managing mobility – measures to reduce the use of the car by discouraging avoidable car use and ensuring that attractive alternatives exist. The target for 2010 is for an increase of 30% in peak hour car use (as compared with 1986) rather than the predicted growth of 70% if no action is taken. Use of public transport at the peak will double and long-distance rail will increase by 50%.

Category 3: Improving environmental quality – emissions by motor vehicles will be significantly reduced and there will be no increase in noise nuisance (1986–2010). Traffic will be concentrated on through routes, road safety will be improved (50% fewer deaths), and the growth in road freight traffic will be reduced. Interestingly, only emissions of nitrogen oxides and unburned hydrocarbons are covered in this category. Other emissions such as carbon dioxide were covered in the National Environmental Plan (NMP+).

Category 4: Support measures – provides the means to achieve the first three objectives through the reforming of transport finance and the applications of the user pays principle; the development of better coordination in transport policy, through better enforcement, through consistent cooperation between transport agencies, and through positive exploitation of market opportunities.

As Gwilliam (1990) comments, the concept of a comprehensive transport plan is one which commands widespread support in the Netherlands, but it is clear that it contains substantial weaknesses. He cites inconsistencies in objectives, the lack of adequate definition of the capability of instruments, the separation of structure planning and financial planning, and the political vulnerability of end state planning. However, it does act as a focus and framework within which to structure more detailed analysis and policy-making, and it is consistent with the targets set in the National Environmental Plan (*Nationaal Milieubeleidsplan*) and its extended version (NMP+).

The Netherlands is perhaps unique in the way it combines transport, planning and the environment. This was apparent in the Fourth National Physical Planning Report and its supplement (VINEX) where location, intensity of development and timing became the main issues for future growth management (Faludi and Van der Valk, 1994). The main theme was that over the next 20 years, growth should be directed wherever possible to existing urban areas or to locations adjacent to existing built-up areas, so that the policy objective of the compact city could be achieved. Most recently in 2001, the Fifth National Physical Planning Report has been released with a new strategy that combines the interests in compactness with the necessity to maintain the Netherland's role as a major international trading nation. The concept of the Delta Metropole or a network city combines Amsterdam, Rotterdam, The Hague and Utrecht into an urban agglomeration of European significance (Bontje and Jolles, 2000). In this Delta Metropole, there would be close links between the constituent parts, based on a strong public transport network. There is a move away from local networks based on the bicycle to regional networks based on trains, trams and buses.

Within this national framework of transport, environmental and land-use plans, significant amounts of detailed research have been carried out, often using disaggregate analysis methods and novel forms of data collection (for example panel surveys and the mobility scanner). The strength of strategic planning in the Netherlands seems to be at odds with other trends which have taken place in policy-making including less government intervention, the desire to reduce public expenditure, and the increasing levels of international integration. Instead of moving towards allowing the market to determine levels of demand and prices, policy in the Netherlands seems to be questioning the broader issues of location, the increases in mobility and the negative external effects of the car. Financial support is being used to fund new infrastructure rather than subsidizing existing services, and new approaches to public management are being sought.

It seems that the Netherlands has reached a decision point where all the negative impacts of the car and high levels of mobility have become apparent within a small country – congestion and a lack of capacity, heavy subsidy in public transport, urban deconcentration, its position as a major distribution centre for northern Europe, environmental pollution, the role of telematics and broad band telecommunications, and high-speed rail connections. The recent changes in policy towards strategic planning reflect both the severity of the problems and the means by which they can be addressed through a mixture of professional pragmatism and scientific research (Priemus and Nijkamp, 1992). At the local level the same thinking has been reflected in creating a clear distinction between spaces that are allocated to general traffic and those where people have priority. The Dutch helped pioneer the *Woonerf* where vehicular access is restricted, initially in older streets but more recently in new developments. This concept was first conceived in Delft, then elaborated in Denmark and Germany, and it is in common use in northern Europe (Ramsey, 1997).

Global warming is also of key concern in the Netherlands and has been used to determine both national and local action in transport (Kroon, 1997). The expected growth in greenhouse gas emissions in the Netherlands is about 20% (1990–2010), but the transport figure is over 30% (Table 7.4). The Netherlands target for CO_2 reduction set at the Kyoto agreement is 6%, so substantial action is required. The overall target is a reduction of about 5 million tonnes in CO_2 emission levels to 2010, of which about a half will come from joint implementation through emissions trading and clean development mechanisms aimed at foreign parties. The other half forms the internal reduction obligation, which has become the responsibility of different ministries (for example, Environment, Economic Affairs and Transport). The alternative options were assessed on a set of criteria, including their potential contribution to the target, the costs for the end-user, the means to monitor and control measures, the timing of implementation, the potential public and political support, and any side-effects. The resultant policy document on climate policy (VROM, 1999) summarized the measures to be introduced. The transport measures included increases in fuel prices, energy labelling on cars, a feebate system on car purchase tax, the enforcement of lower speed limits for cars and lorries, the use of technology in vehicles to maintain efficient fuel use, road pricing, avoidance of short car trips, and fiscal measures to discourage car use in commuting. It is clear that the Netherlands has taken its obligations seriously, and has identified cost effective means to

Table 7.4 Greenhouse gas emissions in the Netherlands

	1990	1995	2010
Transport	31	34	40
Other sectors	186	200	219
Total	217	234	259

Notes
Units in million tonnes CO_2 equivalents
The 2010 figures are business as usual, based on a relatively high economic growth projection
Source: Annema and Van Wee (2000)

reduce the use of the car, and to maintain a consistency in transport planning and policy that matches travel requirements with those of the environment, and the necessity to develop at high densities.

The fiscal measures will mean a real increase of 50% in fuel prices, which in turn will affect the fuel efficiency of the vehicle stock. Tax concessions available at present on long-distance commuting by car (over 30 km) would also be removed. Industries and offices are encouraged to make use of shuttle services with private sector coaches. A further possibility is a doubling of real parking tariffs (Rietveld, 1993). Increases in the length of the highway network will be limited, but the capacity will be increased by 30% through additional lanes, and the use of telematics will give an extra 15% capacity per lane. Priority will also be given to public transport, car pooling, and bicycle use (bicycles still account for 28% of all trips in the Netherlands). Physical measures include strict restrictions on parking spaces for firms and the requirement for firms to develop trip reduction plans as in the United States.

Comment

Although these measures set new standards for environmental and transport policy objectives, achievement of the targets presents real problems as links have to be made with the general economic development of the country. Since 1986, the Netherlands economy has followed the high growth scenario, and it is only more recently that a modest growth rate has been achieved. Much faith has been placed in the SVV-II, VINEX and NMP+ strategic models, and it is unclear whether the policy measures will result in the expected reductions in the use of resources and levels of emissions. The actual responses of individuals may be to accept higher prices so that they can continue to use their cars. It also depends on the government implementing policies which may be unpopular. Road pricing in the Randstad is one of the main instruments for reducing levels of congestion and use of resources in transport, but it is unlikely to be introduced as strong resistance has been roused by various interest groups and political parties.

Road pricing has been rejected by the Second Chamber in Parliament.

> If road pricing would only be located in Randstad, it would create a negative location factor for that part of the country and would increase the trend to disperse industry and housing. Incidently, the social costs of the Randstad are also the highest in the country, which could conceivably justify higher variable costs for automobile use. (Priemus and Nijkamp, 1992, p. 185)

Another debate in the Netherlands is that over the future of Schiphol Airport, which is rapidly reaching capacity. In 1993, the Dutch Parliament decided that the maximum number of passengers should be 44 million a year, a level that would be reached in 2015. It was also agreed that a fifth runway should be built and the possibility of a completely new airport on reclaimed land in the North Sea was raised. Forecasts now suggest that the limit on Schiphol would be reached in 2004, so a public participation exercise was started on the need for the new airport. This process was partly in response to criticism that such a discussion had not been previously initiated, but it was also felt that the divergence of views needed to be exercised (Voogd, 2001). Not surprisingly, the conclusion from this process (December 1997) was that selective and continued growth should be focused on the existing airport and that a further investigation should be initiated on the new airport possibility. The short-term future seems to be that the existing airport will have to meet demand through technical adaptations in control systems and better use of current runways.

It seems that although much analysis has been carried out and a range of major policy measures has been proposed and agreed in principle, the reality is more problematical, particularly if it is seen to be incompatible with the central government's intention to reduce its substantial budgetary deficit. The introduction of such measures, particularly in a recession would make the Dutch economy less competitive. In all countries, when the economic imperative is overriding, all significant policies that might actually reduce levels of growth of mobility and lead to environmental improvements become of secondary importance.

European approaches to evaluation in transport

In most EU countries, the transport infrastructure is publicly owned and so the concern is with a wide range of welfare objectives, rather than a more narrow set of commercial objectives. In addition to economic efficiency (mainly measured through social cost benefit analysis), there are concerns over the environmental impacts of transport (mainly measured through EIA – see Chapter 3) and other broader impacts such as accessibility, finance, and regional development objectives. Although CBA is widely used, many countries also use multi-criteria analysis (MCA), which involves an objectives-led framework that uses indicators for objectives achievement and weights (either assigned by politicians or experts) for the importance of the indicators. From these indicators and weights, a total score for each project can be obtained, giving an overall ranking to assist the decision-maker (see Chapter 3). For private sector projects, the evaluation objectives are different, as the financial assessment carried out on the rates of return on the investment are supplemented by other externally imposed requirements relating to the environment (EIA) and other social objectives.

Given this basic situation, there are important differences in approaches to transport evaluation, which forms a crucial part of project assessment. In Germany, the Federal Transport Plan (*Bundesverkehrwegeplan*) requires a benefit cost study for all projects over 500,000 DM. In this evaluation the alternatives are ranked by three criteria: traffic and congestion effects, regional policy objectives, and other effects such as the elimination of accident blackspots and the international importance of the project. Where possible all costs and benefits are presented in monetary values and results are presented as benefit cost ratios and as net present values. The quantification covers changes in user costs (operating and time costs), infrastructure costs and some external costs (accidents, noise and air pollution). The discount rate recommended is not the market rate, but a rate of 3% which is chosen on the basis of the expected national economic growth rate, in real terms. The basic procedure has not changed since the 1970s, but is now being reviewed (Rothengatter, 2000).

Demand forecasting in Germany takes a range of values and uncertainty is covered by using less optimistic demand forecasts. Regional impact assessments on structurally weak and peripheral areas, together with environmental compatibility assessments, are all part of the standard procedures. In response to the EU Directive on Environmental Impact Statements, an estimation of ecological risks along the proposed route is now investigated, and guidelines for risk estimation have been produced (Wagner and Kleinschmidt, 1995). The environmental compatibility study defines the broad area for investigation, the planning of route alignments, and a detailed explanation of the advantages and disadvantages of each.

In the current review, it seems that a strategic element will be added to the project evaluation to cover the international and regional effects. This means that greater emphasis will be given to the spatial impacts to incorporate the corridor and network effects, and to the environmental impacts to cover effects on biodiversity and habitat (Rothengatter, 2000). This new strategic evaluation will supplement the current project based evaluation, giving a framework within which the more detailed evaluation can be embedded.

Road evaluation in France switched from the primary concern of returns on capital towards a broader based multi-criteria approach. The Directive for Road Transport Evaluation (1986) uses nine categories for evaluation (Table 7.5) of both monetary and non-monetary factors which are then brought together in a series of matrices that rank the options and demonstrate their positive and negative characteristics. The flexibility introduced by such a procedure seems to give the impression that decisions are taken without recourse to formal procedures. Evaluation is just one factor in decision-making and facilitates discussion between the major parties concerned. As with Germany's approach to evaluation, public hearings seem to have only an indirect influence on the planning of roads and other projects as there is no formal presentation of the results of the evaluation to the public. Often the route is decided and individuals have to resort to the courts to change that decision. Even then the decision is unlikely to be reversed, but minor changes including environmental improvements may be made. This means that the time taken from project inception to completion can be considerably reduced. The lead time is often only five years when in Britain it takes over 15 years on average to complete all the procedures (see Chapter 3).

Table 7.5 Criteria used in the evaluation of road proposals in France

1 **Regional and local development** – ease of movement, indirect employment effects, changes in attractiveness of an area and the balance between developed and less developed areas.
2 **Safety** – reductions in numbers of deaths and severe injuries.
3 **Environment and quality of life** – effects on natural resources and ecosystems, human activities including urban planning and access, and quality of life.
4 **Minimization of severe problems** – alleviation of particular problems such as blackspots, congestion and risks from natural phenomena.
5 **Impacts on other transport modes** – switches between modes including gains and losses of revenue.
6 **Direct effects on employment** – jobs created in the construction and maintenance of the project.
7 **Energy and balance of payments** – net change in energy consumption, tourism and taxation.
8 **Public accounts and accounts of concessionaires** – net change in expenditure and receipts for public expenditure and the ability for private operators to recoup costs.
9 **Cost benefit assessment** – all possible factors are included to give a net present value and first year rate of return.

In both France and Britain the importance of development objectives is considerable. In France, DATAR (*Délégation à l'Aménagement du Territoire et à l'Action Régionale*) argued the case that the original Master Plan for Roads (1960) favoured the prosperous central areas as it was based on the predicted continuation of existing trends which in turn suggested that the greatest economic rates of return would take place in these core areas. The DATAR plan used criteria for national and regional development to promote greater investment in peripheral regions. However, the government was also concerned about restricting the levels of public funding of the motorway systems, and was anxious to encourage private concessions for toll road construction. This problem has been augmented by the move towards decentralization of power in France. The regions can now use their powers, their considerable financial resources, and political will to favour free motorways or roads of an equivalent quality. But the central government now favours public funding for public transport (especially rail), and private funding for major road schemes (Grandjean and Henry, 1984).

Table 7.6 summarizes the approaches to road and public transport evaluation used in Britain, Germany, France and the Netherlands. The formal use of cost benefit analysis which keeps the power in the hands of the planners is more likely to be found in Britain and Germany. Multi-criteria analysis gives

Table 7.6 Evaluation methods used in the four countries

	United Kingdom	Germany	France	Netherlands
Direct impacts				
Construction costs	CBA	CBA	CBA	CBA/MCA
Disruption costs	CBA			
Land and property costs	CBA	CBA		
Maintenance costs	CBA	CBA		CBA/MCA
Operating costs		CBA		CBA/MCA
Vehicle operating costs	CBA	CBA	CBA	CBA/MCA
Revenues				
Passenger cost savings				
Time savings	CBA	CBA	CBA	CBA/MCA
Safety	CBA	CBA	CBA	CBA/MCA
Service level				
Information				
Enforcement				
Financing/taxation				
Environmental impacts				
Noise	MI	CBA	CBA	MI/MCA
Vibration	MI			
Air pollution – local	MI/QA	CBA	CBA	MI/MCA
Air pollution – global	MI		CBA	
Severance	MI	CBA		QA
Visual intrusion	MI/QA			
Loss of important sites	MI		QA	QA
Resource consumption				QA
Landscape	MI	QA		QA
Ground/water pollution	MI			QA
Socio-economic impacts				
Land use	QA			QA/MCA
Economic development		CBA	QA	QA/MCA
Employment		CBA	QA	
Economic and social cohesion				
International traffic		CBA	QA	
Interoperability				
Regional policy	QA	CBA		QA/MCA
Conformity to sector plans				
Peripherality/distribution			QA	

Notes

CBA = cost benefit analysis with monetized impacts

MI   = measured impacts

QA   = qualitative assessment

MCA = multi-criteria analysis

The United Kingdom entries cover road transport only, but the other countries cover all transport modes.

Source: Based on Bristow and Nellthorp (2000)

the decision-makers more power and responsibilities to weight goals and other criteria, and this procedure is more likely to be used in France for road evaluation and in the Netherlands more generally. Other forms of evaluation, such as the use of participatory methods, which involve the use of softer qualitative techniques and extensive debate with the people affected, seem to have been ignored. In Britain, the financial and economic factors dominate evaluation, but in the other three European countries regional policy objectives and environmental factors add to the financial and economic criteria.

This Chapter has placed the experience from the United Kingdom within the broader context of transport planning in the EU and in particular in three key European countries. Although many of the problems are similar, the different traditions have led to different approaches to analysis and different solutions to the problems presented. The experience in the United States is also illuminating in this respect, and in the next section some of the main elements of their transport and planning processes are introduced.

Transport planning in the United States

Conventional thinking in the 1950s and 1960s was that roads should be built to accommodate car traffic in the era of universal motorization. Car ownership in all advanced industrial countries was rapidly increasing and saturation levels would be reached where some 90% of all adults (17–74 years of age) would have their own car. This would give a car ownership level of 650 cars per 1000 population. This figure for vehicle ownership in the United States was exceeded in 1984, and in 2000 there were over 800 vehicles per 1000 population. The challenge for highway planners was to develop a network of all-weather highways to meet that expected demand. In the US, a series of standard procedures was developed for application to all urban areas — *Better Transportation for Your City* (1958).

The methods used were aggregate in scale:

♦   Fratar method for trip generation;
♦   gravity models for trip distribution;
♦   traffic diversion curves, based on empirically derived travel time ratios;
♦   shortest path assignment algorithms.

They were designed to place urban transport planning on a more systematic basis. Several seminal studies were carried out, including the Detroit Metropolitan Area Traffic Study (1953–55), where many of the calculations were done by hand, and the Chicago Area Transportation Study (1955). These studies were a tribute to the pioneering spirit of the time, and the enterprise of Douglas Carroll who directed both of them. The optimistic view of that time is captured in the two benchmark books by Meyer, Kain and Wohl (1965) and Creighton (1970) on urban transportation planning.

In March 1962, a joint report to the President on urban mass transportation stated:

> transportation is one of the key factors in shaping our cities. As our communities increasingly undertake deliberate measures to guide their development and renewal, we must be sure that transportation planning and construction are integral parts of general development planning and programming. (Weiner, 1985)

The importance of this statement is that it marked a switch of resources from highway construction to urban mass transportation, and there was a requirement (in the Federal Aid Highway Act, 1962) to develop programmes which properly coordinated and evaluated the impact of future development of the urban area. The Bureau of Public Roads set up the basic elements of the '3C Planning Process' (continuing, comprehensive, cooperative) for which inventories and analyses were required to cover:

- economic factors affecting development;
- population;
- land use;
- transportation facilities including those for mass transportation;
- travel patterns;
- terminal and transfer facilities;
- traffic control features;
- zoning ordinances, subdivision regulations, building codes etc;
- financial resources;
- social and community value factors, such as preservation of open space, parks and recreational facilities; preservation of historic sites and buildings; environmental amenities; and aesthetics.

The 3C Planning Process led to the development of standardized procedures, computer software and training courses for transport professionals, and by 1965 all 224 urban areas had started their urban transport planning process (Weiner, 1985). For the first time federal grants were available for mass transit construction costs, and for the acquisition of facilities and equipment (Urban Mass Transportation Act, 1964) provided that those costs could not be paid for out of revenues.

It was also during this period in the turbulent 1960s that two other issues came to prominence. Firstly, there was a reaction against the highway construction programmes, initially in San Francisco, but soon mirrored across the Western world. The premise on which highway planning was predicated, namely that the demand for car travel had to be met by increased road capacity, was shown to be false. The impact of the Buchanan Report in Britain

(Chapters 2 and 3) has already been noted. The equivalent report in Germany (Hollatz and Tamms, 1965) argued that the growth in car ownership and use had to be met through the construction of orbital and ring roads, combined with restraint on car use in city centres and heavy investment in public transport. In the United States, Meyer, Kain and Wohl (1965) tested a series of hypotheses through a range of exhaustive empirical investigations. These hypotheses were:

- the decline in city populations and densities may be attributed to poor public transport;
- a rail-based solution would be cheaper than a road based one;
- travellers would be willing to switch to public transport if the quality was improved;
- urban transport shapes the city.

In the 20 years after the war, public transport use in the United States declined by 64% (1945–1963), but route miles of railway increased by 2% and route miles by bus increased by 30%. The situation in Britain was slightly different. Here, peak public transport use was in 1952, not in 1945. The decline over the subsequent 30 years was about 30%, with route miles also being cut (by 20%). The conclusion reached is that if high levels of mobility are desirable and car ownership continues to rise, then public policy should be concerned with the construction of extensive networks of new roads in urban areas and in investment in public transport so that travellers have a real choice of alternatives.

The second issue, again picked up in the Meyer, Kain and Wohl book (1965), was the lack of coordination between different government agencies, principally relating to housing and transport. The Department of Transportation was created (1966) to match the new Department of Housing and Urban Development (1965). The end of the 1960s marked a period of uncertainty in the use of the systems approach to transport analysis as models had not matched expectations, particularly at the local level. It was also realized that the large-scale aggregate approach, together with the well established procedures for data collection and analysis, did not cover the broader social goals or behavioural factors.

Different approaches seem to have been adopted by Buchanan and Meyer, Kain and Wohl. *Traffic in Towns* attempts to examine a range of options through a series of practical studies that represent the spectrum of urban areas in Britain. The American study of *The Urban Transportation Problem* is not so comprehensive and focuses on a set of particular issues in a systematic manner, such as the relationship between housing and transport, and race and transport. Perhaps the most important contrast is that Buchanan's options examine different levels of redevelopment and the notion of environmental

areas around which traffic would be diverted, whilst the focus in Meyer, Kain
and Wohl is on the modal split question and the balance between public and
private transport. Both issues are encompassed in Appleyard's concept of
'liveability' and environmental capacity (Appleyard, 1981) where it is
suggested that there is an inverse correlation between traffic volumes and
neighbourhood (liveability). The essence of the argument is to make the car
more amenable to the city and *not* the city to the car. Appleyard (1981, pp.
130–31) claims that the concern in *Traffic in Towns* for the environmental
impacts of traffic predate official American interest in the problem by about
five years.

On both sides of the Atlantic, the early 1970s marked a watershed in
thinking as high oil prices, global recession and increased unemployment
caused political and economic uncertainty. Priorities switched from road
construction to the means to improve capacity and safety through traffic
operations (TOPICS), and the requirement to hold public hearings on
proposed highway projects. The continuing urban transport planning process
should cover five elements (Altshuler, 1979):

- surveillance – monitoring change;
- reappraisal – review and update proposals;
- service – assistance in implementation;
- procedural development – improved analytical skills;
- annual report – wider communication with the public and officials.

Even though thinking on the environment may have come later to the
United States, action has been far more decisive than in Britain. The
pioneering National Environmental Policy Act (1969) required environmental
impact statements for major federal actions and for all legislation, and this in
turn necessitated a systematic interdisciplinary approach to planning and
decision-making. The Council on Environmental Quality would implement
the policy. It was only in June 1985 that the European Community issued the
Directive on the Assessment of the Effects of Certain Public and Private
Projects on the Environment, including the construction of motorways and
express roads – interpreted in Britain as trunk roads over 10 km in length
(Article 4(1)), and other roads or urban development projects at the discretion
of individual member states (Article 4(2)). The Directive took effect from July
1988 (Chapter 3).

The 1970s became the decade of public transport, with significant increases
in capital and revenue support. The Urban Mass Transportation Assistance
Act (1970) gave priority to schemes for helping elderly and handicapped
travellers whilst all highway projects had to give a full analysis of economic,
social and environmental effects (Federal Aid Highway Act, 1970). The main
innovation in analysis methods was the use of behavioural and disaggregate

models, together with a realization that short-term, quick-response, simple methods were most informative for decision-makers. The switching of resources continued with the Federal Aid Highway Act (1973) which permitted the use of highway funds for urban public transport for the first time. A year later, the national Mass Transportation Assistance Act allowed the use of federal funds for public transport operating subsidy.

San Francisco's Bay Area Rapid Transit (BART) epitomizes many of the debates concerning public transport. Operations began in October 1974 over the 71 mile (114 km) system, and by 1976 BART was carrying only half the forecast passengers and the operating loss had reached $40 million. These losses compounded the high capital cost overrun of 150% and the high operating costs, some 475% over forecast. Average costs per ride were twice those of the bus and 50% greater than a standard American car (Webber, 1976). One of the main arguments for the investment was the potential for BART to attract travellers from the car, but with only 5% of the peak trips, it has made little difference. The importance of access time to and from the system was underestimated. Webber (1976) concludes

> it is the door-to-door, no-wait, no-transfer features of the automobile that, by eliminating access time, make private cars so attractive to commuters – not its top speed. BART offers just the opposite set of features to the commuting motorist, sacrificing just the ones he values most.

BART has not noticeably influenced land-use patterns, except in particular locations such as at Walnut Creek (about 10 miles east of central San Francisco). The intention was to create highly accessible locations which would prove attractive to new businesses, offices and housing developments. The localization effects would generate further concentration at these sites as multiplier effects took place. The mistake was the expectation that the railway would make any impact on accessibility in an urban area which already has a highly developed road network (Cervero and Landis, 1995). Even where there were opportunities to develop 'transit villages' around BART stations, there was local opposition from residents, and there was also a difficulty in deciding whether priority for more development or more parking at stations was a better way to make up the shortfall in BART patronage (Knack, 1995). The technical planning process set no formal goals, looked at no alternatives, and made no formal evaluation until late in the process. The problem was to alleviate traffic congestion coupled with the protests over freeway construction, and the mechanisms for public consultation were weak (Hall, 1980).

The third factor was the availability of federal capital to match locally raised capital. With the setting up of the Urban Mass Transportation Administration within the Department of Transportation (1968) and the subsequent commitment of federal aid to public transport investments, huge new investment projects could now be undertaken. The funding was concentrated in 1600 miles (2500 km) of new urban rail lines including Washington, Los Angeles, Baltimore, Detroit and Atlanta. It seemed that during the 1970s there was a genuine feeling supported by local politicians that rapid transit systems could get people to switch from car to rail. New technology and heavy investment in urban rail would provide the solution to urban congestion.

In retrospect, the arguments may seem naive, but given the range of possible alternatives there may have been little option. The path followed in Britain was somewhat different. Heavy capital investment in roads was curtailed in the 1970s, but there was little diversion of resources to capital investment in urban rail systems – with the exception of the Tyne and Wear Metro in Newcastle (1974–1984), the extension of the Victoria line in London (1971), the extension of the Piccadilly line to Heathrow (1977), and the new Jubilee line in London (1979). In both countries, the short-term response was a heavy commitment to further subsidy for both bus and rail systems.

Transport planning in the United States in the 1980s moved towards the managing, maintenance and replacement of existing facilities, together with an extensive analysis of the means to limit the use of the car in the city. Apart from the problems of physical capacity, there was an increasing concern over energy consumption in transport and the growth in transport-related levels of pollution. All these factors helped focus attention on the car. Urban policies were also directed at the means by which city centre decline could be reversed. The confidence of long-term planning, characteristic of the previous two decades, was replaced by a more cautious pragmatic incremental approach to planning.

The key to this new approach has been the decentralization of decision-making and a reduction in federal intrusion into local decisions (Weiner, 1985). The complexities of planning as a social process with important environmental and ecological dimensions, together with tighter fiscal constraints and increased private sector involvement, all make it more appropriate for devolved decisions to be taken. The rigid large-scale physical approach to planning was not suitable for smaller-scale decisions which required local flexibility. Coupled with this decentralization was a necessity to reinvest in upgrading the existing road infrastructure. The Surface Transportation Assistance Act (1982) increased the '4R Programme' – resurfacing, restoration, rehabilitation, reconstruction – by raising user charges from 4 cents to 9 cents a gallon of petrol. The new Urban Transport Planning Regulations (1982) retained the requirement for a Transport Improvement Program (TIP) including a Unified Planning Work Program (UPWP) for all areas with a population over 200,000, but the process was to be self-certified by the state and the Metropolitan Planning Organizations (MPOs).

However, at the local level there is a separation of transport and planning functions, and a shortage of available expertise. The land development process is mainly a private sector operation and there is little expectation that a plan will be implemented. Even zoning is carried out to meet expectations rather than reality, and much planning is merely modifying existing plans or carrying out rezonings (Deakin, 1990a). Local government had little responsibility for transport and was mainly concerned with land-use allocations. Similarly, the state and regional agencies were not responsible for land use. There was a fundamental flaw with the system, and development rights were granted in the understanding that if one local authority did not do it, employment would move elsewhere (Cullingworth, 1999).

Underlying all transport planning problems of the 1980s and early 1990s is the key problem of congestion and the prospect of total gridlock in city centres and suburbs. Demand patterns are now so complex and suburbs are no longer dormitory locations for commuters working in the city centres. Suburbs in the main metropolitan areas are now the places where shopping centres are located and the recent growth has been in business parks and the 'horizontal' office blocks (groundscrapers). They are attractors and generators of traffic, yet the evidence on whether trip distances have increased as a result of suburbanization is less than clear (Cervero and Hall, 1990; Gordon and Richardson, 1989). The Californian Transportation Commission (1987) estimated that congestion results in 75 million hours lost annually within the State, and about half of this delay is caused by non-recurring incidents such as accidents, lane closures and unpredictable events. Building new roads or expanding existing ones would not solve that problem. Downs's concept of 'triple convergence' means that there is first a switching effect (spatial convergence) as motorists use the new road, secondly other motorists will travel at more convenient times on the road (time convergence), and thirdly travellers will switch from public transport to driving on the new road (modal convergence). The total effect is that a new congested equilibrium will be produced (Downs, 1992).

Transportation Systems Management (TSM) has been extensively used to shift demand over space and time through more flexible working practices and through the use of vanpools and carpools. With the cutbacks in the federal budgets, the private sector has become more involved in funding roads, with its contribution reaching about 20% (Vuchic, 1999 and TRB, 2001). Projects range from local improvements to a new generation of toll roads financed by bonds and toll revenues. Traffic volumes of 50,000 vehicles a day are needed to justify toll financing so that interest charges and maintenance costs can be paid (Deakin, 1990b). Federal aid is now also available at a 35% matching level for a small number of toll road demonstration projects, but it is often these supplementary funds and other guarantees from developers and government (for example, development impact fees and tax benefits) which make the decision worthwhile. There are few new opportunities for toll roads in the United States as all main possibilities have already been built, and it is only where development rights are made available that the private sector will be interested in participating (Gomez-Ibanez and Meyer, 1995). Roads built to relieve congestion require more complicated packages as costs are higher (as construction is often in urban areas), and there is likely to be greater opposition. Consequently, private sector interest is lower. The main role for the private sector may be as innovators, where they set benchmark standards against which the performance of the public bodies can be measured and stimulated, either through innovative charging methods or through the encouragement of new forms of high occupancy vehicle priority and charging.

Other forms of road pricing seem less attractive in the United States and there is little evidence of peak pricing except on the Washington Metro and on certain bridges (for example, San Francisco Bay Bridge) or tunnels (for example, those going into Manhattan). But even here, there is only an implicit acceptance of pricing as high occupancy vehicles (HOV) are exempt and also there is no difference in rates between the peak and off-peak. Parking is usually provided at a low cost except in the central core (90% of US commuters do not pay for parking), and zoning standards often overestimate the amount of parking that developers should provide. It is likely that there will be a huge resistance to road pricing in the United States. Comprehensive regional planning seems as far away as when the Regional Planning Association of America called for it in the 1920s (Cervero and Hall, 1990).

Superimposed on the threat of gridlock is that of the environment. It is here that some of the most interesting innovations have taken place as land use is being controlled to reduce the rate of growth in traffic. In Southern California, Proposition U has reduced the allowable development on most land zoned for commercial development. The measure halved allowable development along certain main roads from 3:1 to 1.5:1 (the floor area ratio). The Traffic Reduction and Improvement Program (TRIP) allows the Council in Los Angeles City to designate any neighbourhood a 'traffic impact area'. For each area of transport, a specific plan is developed including the requirement for the developer to calculate the trips generated by the proposed development (from the *Manual of Trip Generation Rates* produced by the Institution of Transportation Engineers). The developer must then mitigate the effects of those trips by reducing the afternoon peak hour trip generation of the project by at least 15%. Failure to achieve this reduction results in a fee being paid into a trust fund and the monies can be used to pay for improvements in the impact area's transport plan (Wachs, 1990). The introduction of parking charges at workplaces has been the most effective way to meet these targets with wages being raised to compensate for the loss incurred in paying for workplace parking.

Regulation XV imposed by the Southern California Air Quality Management District (SCAQMD) requires that each employer with 100 or more employees has to ensure that its workforce achieves an Average Vehicle Ridership (AVR) threshold for journeys to work. The AVR is the vehicle occupancy rate and relates to all employees and all cars. Thus the impact of the use of public transport, bicycle and walk modes will all be positive. The actual levels of AVR depends on the city location with the Los Angeles CBD value being 1.75 and the suburban value being 1.3. Again, fines are imposed on non-achievement of the targets, and it was estimated (Atkins, 1992) that in August 1991 some 209 employers out of a total of 6,900 were in violation. In each of these cases, the problem has been addressed through the workplace, either in terms of reducing the impact of new developments or in reducing the numbers of cars used for the journey to work. However, the environmental and congestion problems created by the car do not only relate to the work journey, but to a multitude of other activities which are much harder to control (for example, social and recreational trips). Work journeys in both the United States and Britain are relatively stable in number, and form a diminishing proportion of the expanding travel market.

Although the United States has thought about road pricing, it has made even less impact than in Europe where it is now seen as providing one means to reduce the use of the car in cities. The additional problem for the United States is that many of the cities are already very low density, with no effective choice of travel mode. So even if road pricing were acceptable, its impact would be likely to be low as drivers would have to continue to use the car. The advice from Downs is to get politically involved and learn to enjoy congestion:

> Get a comfortable, air-conditioned car with a stereo radio, a tape player, a telephone, perhaps a fax machine, and commute with someone who is really attractive. Then regard the moments spent stuck in traffic simply as an addition to leisure time. (Downs, 1992, p. 164)

In summary, the system needed changing as there were administrative, financial and political difficulties with transport and planning decisions as they had evolved over the previous 30 years. Table 7.7 gives a summary of the complexity and conflicting objectives for metropolitan planning as they existed in 1990.

The Intermodal Surface Transportation Efficiency Act (ISTEA – commonly pronounced 'iced-tea') was passed in 1991 and tackled the three issues of congestion, air quality and finance. ISTEA requires the production of long range plans (LRP – 20 year time horizon) and transportation improvement programmes (TIP – 3 year time horizon) which, in the metropolitan areas,

Table 7.7   The 15 metropolitan planning requirements

1	Preservation of existing facilities and more efficient use of existing facilities.
2	Consistency of transportation planning with energy conservation programmes.
3	Relief of congestion – now and in the future.
4	Effect of transportation on land use.
5	Programming of enhancement activities.
6	Effects of all transportation projects – even if not federally funded.
7	International borders, intermodal facilities, parks, historic sites.
8	Connectivity of metropolitan area roads with roads outside the metropolitan area.
9	Transportation needs identified through management systems – highway pavement, bridge, highway safety, traffic congestion, public transportation facilities, and intermodal transportation facilities.
10	Preservation of rights of way for future projects – identification of corridors.
11	Methods to enhance movements of freight.
12	Use of life cycle costs – bridges, tunnels, pavement.
13	Overall social, environmental, economic and energy effects.
14	Expansion, enhancement of transportation services.
15	Capital investments for increased security in transit systems.

Source: Paaswell (1995)

would be carried out by the Metropolitan Planning Organizations (MPO). The argument behind ISTEA seems to have been that it was now time to bring together all the different forms of funding that historically had favoured road building into a new integrated system where federal grants (80%) could be matched by local grants (20%). It was recognized that some 100 US urban areas violated national standards for permissable levels of ozone, carbon monoxide and particulate emissions. Part of the aim was to establish what each urban area was actually doing to meet NAAQS (National Ambient Air Quality Standards) by specific dates (Paaswell, 1995).

Three main groups of measures were being encouraged. First, Transportation management areas (TMAs) were designated for cities over 200,000 that did not meet NAAQS targets. Congestion management systems (CMS) had to be established with clear objectives and benchmarks to congestion mitigation. Secondly, employee trip reduction programmes were introduced for all companies employing more than 100 people, so that vehicle occupancy would increase by 25% in the peak hour, as described earlier. Thirdly, and perhaps most importantly, no new highways were to be constructed in non-attainment areas where these would encourage more single occupancy vehicle travel.

Much of the thinking behind ISTEA was environmental, and it strengthens the pioneering NEPA legislation and the subsequent Clean Air Act (and Amendments). It marked a change in thinking away from building new roads to reduce congestion, and it also recognized that insufficient reductions could be achieved through traffic management measures alone. The answer was to look for ways to increase vehicle occupancy and to use demand management alternatives. However, the possibilities of substantially raising the costs of travel through increases in fuel prices, parking charges and road charging were not considered, nor was public transport really seen as an alternative. The crucial links between transport planning and land-use planning do not seem to have been made, as the metropolitan planning organizations have not acquired additional powers; rather there seems to be an expectation that partnerships will develop between the main agencies naturally. This expectation has not been met, as local and city governments have their own agendas and are attempting to gain advantage (Paaswell, 2000). This means that in the Tri State Metropolitan Region, there are three MPOs (New York, New Jersey and Connecticut), which are all competing with each other for development and investment, with no mandate to coordinate or cooperate.

Comment

The basic philosophy behind land-use and transport planning in the United States is very similar to that in Britain. In the early stages with the development of the systems approach, Britain followed the theoretical and analytical path pioneered in the United States, but more recently with the increased use of disaggregate and behavioural methods that interaction has been more balanced. In both countries there is at present a pause in the development of methods and their application in planning practice. This pause is in part due to the move away from large-scale strategic planning, but more particularly to the growing interest in smaller-scale localized application of particular analysis methods to immediate problems. The concern over comprehensiveness and order is being replaced by one of selectivity and specialization.

The main difference between approaches has been the trend in the United States towards decentralization of power, whilst in Britain increased centralization of power and resources has taken place. In the United States, the states have always had more power over local decision-making and the raising of local taxes than have the counties in Britain. The 1980s were characterized by reductions in public budgets at both the national and the local levels, yet in the United States there is more opportunity to raise cash locally. Tax hypothecation is permitted in the United States through the levy on petrol prices, and the private sector seems to have been more actively involved in the new generation of toll road construction. The planning system in Britain is still more restrictive than that in the United States, and as such the opportunity for private investment in roads has been limited. The possibility of federal support for capital expenditure in rapid transit systems and toll roads, together with advantageous development benefits, have all resulted in greater private sector activity in the United States (Deakin, 1990*b*).

Comparable schemes in Britain are more limited and are much more closely linked in with the business cycle in the building and construction industries.

The other main difference between the two countries has been the continued subsidization of public transport in the United States, whilst in Britain the government has managed to reduce subsidy levels, principally through regulatory reform in the bus industry and through the imposition of severe productivity targets for the railways. Transport professionals have never really been at ease with the increased politicization of transport issues. As Wachs (1985a) comments, subsidy is commonly seen as inefficient and analysis has been 'so uniformly critical of American public transit policy for so long, yet the trend towards greater subsidies, more inefficient fares structures, and continuing shifts in service towards the lower density, higher cost markets continues'. Many of the studies into the reasons for subsidizing public transport have been ignored as the technical arguments do not address the political criteria which are dominant in the process of resource allocation (TRB, 2001). Equity, in political terms, is the share that constituency gets of the budget and not the extent to which those who benefit from a service actually pay for it or actually need it. Efficiency is the quantity of service produced by each dollar of subsidy, not the fare box revenues or measures of productivity. The gasoline tax seems to summarize the difference between political and technical decision-making. The gasoline tax is a user tax, yet 20% is allocated to transit funding, but the majority of Americans never use buses. So the tax is in effect being paid by motorists for the non-use of buses.

In the 1990s, the policy and planning agendas in Britain and the United States converged, as issues relating to congestion, the environment and finance came to the forefront, in the United States through ISTEA and in Britain through a variety of studies but no coherent policy initiative. There also seems to have been an agreement over what should not be done, but less accord on the actions to take. Both countries were still interested in building more roads in the 1990s to address the congestion issue, but in the United States public transport was not seen as an alternative to the car. The only way to reduce congestion and increase efficiency (and meet air quality targets) was to make better use of the car through higher occupancy. Much of the emphasis was on involving the major employers in that process. Such a strategy has come more recently to Britain, where company transport plans are still relatively new and follow European experience. The other key difference here is Britain's use of pricing mechanisms to get people to reassess their travel patterns and to make greater use of public transport. In the United States, there still seems to be a deep resentment at paying anything near the true costs of transport.

The institutional framework in the two countries are also different. In the UK, there is a close liaison between planning and transport from central government through the regions to the local implementation authorities, with most funding coming from the centre and the regions. Even land-use planning is subject to national guidance. Local transport plans are part of the development plan process, which link across all public services. In the United States there are no such links, as transport issues still reside at the state and regional levels, with the MPOs coordinating federal funding at the local level. The development process is very much a local issue with funding coming from the developers, and areas are seen to be in competition with each other (Transportation Research Board, 2001). The British model is one of coordination and integration between the sectors, whilst in the United States there is still a clear separation of powers and responsibilities.

Conclusions

Transport planning analysis over the last 30 years in all countries investigated here has been dominated by an engineering approach in which the quantitative values (for example, capacity and network expansion) are considered more important than the qualitative factors (for example, safety, distributional issues and externalities). Demand has usually been uncritically accepted as given and the job of the transport planner has been to accommodate that demand, principally through increasing the capacity of the system (mainly the road system) and by implication the use of the car. Trend extrapolation and demand forecasting methods have been used to estimate future growth in demand. Again, the aim of the analysis was to test alternative strategies to meet the expected growth in demand.

At the same time the context within which transport was operating has itself changed, with demographic and other structural changes in society, with increases in leisure time, with changes in lifestyles and the growth in levels of affluence, with shifts in the labour market, with technological developments, and with concern over qualitative factors such as the environment. Public policy-makers in transport were also facing a series of difficult questions. There was a growing backlog in road and rail infrastructure maintenance and reinvestment; there were the insoluble problems of congestion in urban areas; there were unacceptable delays to public transport users on the network and at termini; there were threats to the viability of some public transport services; there was an awareness of incompatibilities between the switch to the car and environmental objectives; and there was a need to accommodate expensive new technologies in vehicles, in information systems, in the distribution of goods, in the organization of transport, and in the management of the system.

Governments were also making an important contribution to the debate. Policy priorities changed significantly in the 1980s with a concern over regulatory reform and the opportunity to increase competition within transport, with cutbacks in public budgets available for both capital investment and revenue support. Superimposed on these changes is the growing internationalization of many transport decisions as policy initiatives were being taken outside of the national arena, principally by the EU, but also through international agreements. Yet most of the available investment funds were still being used for road investment rather than public transport investment.

In short, the position of transport policy as a strict regulator was being questioned for a range of reasons (Noortman, 1988). These included a lack of consistent and non-conflicting objectives; a lack of adequate and effective policy instruments; limited budget capacity to implement policy actions; inertia in transport policy caused by long-lasting bureaucratic procedures; and the lack of a suitable and efficient legal system for a creative and trend setting policy.

The underlying philosophy of the public policy approach was being questioned. The two main issues related to the public goods and the externalities arguments. The public goods argument refers to the indigenous role of transport and infrastructure in society, in which equity considerations and monopolization objectives are as important as efficiency objectives. The externalities argument concerns both the positive aspects, such as stimulating economic and regional development by improving accessibility, and the negative aspects, such as the need to reduce air pollution and noise nuisance (Gent and Nijkamp, 1989). National governments may not be rejecting the public policy arguments on transport, but they are examining ways in which they can reduce the ever increasing financial commitment to transport and other public sector activities.

Superimposed on these two issues is the changing role of government, whether it should be interventionist or whether it should leave the market to establish the appropriate levels of investment and prices charged. In general, governments in the 1980s moved to the right with reductions in public expenditure and the basic strategy of charging the user the full costs of services used. Although there has been a political switch back to the left with more socialist policies, this has not meant a change in direction, and the 1990s saw a continuation of the market approach, but with a stronger welfare element within it. Yet, even here, some governments (for example, UK and US) seemed much more willing to impose the policies of the social market more strongly than others (for example, Germany and France). It is only in the Netherlands that policy and planning priorities have been introduced to integrate transport planning with environmental issues, but even here there are now considerable problems with the implementation, as governments cannot afford to be seen as anti-motorist. Although there has been radical political change over the last 20 years, there seems to have been no major paradigm shift in transport planning, even though substantial change has taken place in policy priorities and detailed analysis approaches.

Chapter 8

Transport agenda 21: the way forward

Introduction

Transport planning has undergone radical change in the last 40 years, and it is now barely recognizable from its origins in the highway building movement and its concerns over increasing the capacity of the system to meet the expected levels of demand. The focus in this Chapter is on the recent past as it relates to passenger transport, with a heavy emphasis on strategic and urban planning issues. It reviews the most recent developments in transport planning (1990–2000), and it describes the new agenda and the key role of transport planners in promoting sustainable development. Over this period, there have been three great changes in the requirements placed on transport planning, namely the huge growth in car ownership and congestion, the withdrawal of the state from the provision of transport services through regulatory reform and privatization, and the new environmental debates.

The 1990s have been remembered for the huge growth in the numbers of cars and drivers. The costs of acquisition and use of the car were significantly reduced with extensive systems of subsidies being provided to many motorists through their employers and through the pricing of transport at levels substantially lower than their full social and environmental costs. Traffic levels in many car dependent countries have doubled (1975–1995), but the expansion of the infrastructure has been more modest, typically a 10–15% increase in the road network (mainly the motorways). These two underlying trends have resulted in the inevitable increase in congestion and this situation is well-illustrated with data from the EU, Japan and the United States (Table 8.1).

The picture given by each of the three major economic powers in the world is very different. The EU scores highest on the number of cars and taxis per unit of road, but this gap is narrowed when all vehicles are included as there are substantially higher numbers of trucks in Japan and the United States. The increase in the number of vehicles per unit of road has been about 15%, somewhat higher in Japan and lower in the United States. Japan has the highest density of vehicles, but makes less use of them. The modal share for car is substantially lower than the levels in the United States and the EU, and the average distance that each person travels is also lower at 75% and 50%

of the EU and United States levels, respectively. Across all three economies, there has been a 22–28% increase in distance travelled and a corresponding increase in the numbers of vehicle-kilometres per person per year. Congestion has become substantially worse over the recent past, and this has in turn led to a reassessment of policy priorities.

Table 8.1 Measures of congestion and traffic in the EU, Japan and the United States

Measure of congestion	1987			1997		
	EU	Japan	USA	EU	Japan	USA
Cars and taxis per km of road	41.4	26.2	22.0	47.8	40.4	20.1
Vehicles per km of road	52.3	60.8	29.5	60.6	72.5	32.8
Vehicle-km per km of road (thousands)	596	483	473	749	601	619
Distance travelled per person per year (km)	9000	6800	14500	11300	8700	17800
Percentage of all distance travelled by car	80.4%	46.9%	95.3%	84.8%	54.2%	95.8%

Notes
Roads include motorways, national roads, state roads and municipal roads.
Not all dates are exactly comparable, but all are within the period 1985–1987 and 1995–1997.
The levels of congestion in Great Britain are substantially higher than those for the EU as a whole (see Table 5.8).

Important conclusions have been reached by most governments in developed countries. Congestion is going to get worse as the capacity of the network will never increase at a level to match the increase in demand. Even if it was possible to invest in expanding the infrastructure, for financial and environmental reasons this is not seen as desirable. There are other higher priorities for public expenditure (for example, health and education), and investment in new roads has often proved to be politically unpopular for many governments. The role of the transport planner has changed from the provider of roads and additional capacity to exploring the means by which the existing capacity can be better used and allocated to priority users (Dunn, 1998).

This links into the second great change. In the past governments have always played a major interventionist role in transport decisions. The underlying philosophy was that transport should be made available to meet the needs of the population and that everyone could expect a minimum level of mobility, almost as a right. To this end, once the basic road network has been established, investment was switched to public transport in the form of capital and revenue expenditure. Very few roads or public transport services were operated by the private sector. This underlying philosophy was common across most countries, perhaps with the exception of the United States where the market was allocated a much stronger role (Roth, 1996). This has all changed as the market based approach has now been extended to many other countries and the role of government has been significantly reduced with market forces determining both the quantity and the quality of transport services.

The new government policy on transport means that services should be provided by the private sector wherever possible, service levels should be determined competitively, not in a coordinated fashion, and fares should be market priced. Coupled with these fundamental changes is a move towards a greater precision in defining the objectives of public transport enterprises, particularly financial performance objectives and quality of service standards. The traditional role of the transport planner has been radically modified as transport services are now provided by transport professionals rather than planners.

The third great change is the environmental impact of transport and the key role that transport should play in the achievement of targets to reduce global warming, to reduce dependence on non-renewable energy sources, and to minimize local pollution and adverse social impacts. In the past, it was accepted that transport had an impact on the local environment through unacceptable levels of road accidents, through increases in noise levels, through community severance, through visual intrusion, through vibration from lorries, and through some pollution effects (for example, carbon monoxide and carbon emissions from the incomplete combustion of fuel). But now, the environmental agenda is much wider and includes global emissions (principally carbon dioxide), an extended list of local emissions (including small particles which cause breathing difficulties), the consumption of non-renewable resources (fuel), the use of land, and the impact on the local ecology and ecosystems (RCEP, 1994).

These three fundamental changes mean that the traditional role of the transport planner must be reinterpreted so that the new requirements can be accommodated. In the next section the changes are further developed through an assessment of the reactions to the new requirements and the means by which the transport planning process has responded.

The new agenda

Transport planning has been completely transformed from an essentially technical activity based on a simple demand led assumption (sometimes called predict and provide[1]) to a much more complex approach that attempts to place limits on mobility through pricing, regulation and other control strategies. The underlying principle is one of demand management. In addition to tackling congestion in cities and along many inter-urban routes, transport planning has to address questions of meeting environmental standards, identifying pollution hotspots, setting and achieving traffic reduction targets, but at the same time ensuring that all people have appropriate levels of accessibility to jobs, services and facilities. The underlying question here is, whether transport planning should do more than just respond to these fundamentally new requirements, and whether it is prepared to take a lead in developing the new agenda.

Transport planning is essential at both the strategic and local levels, but the nature of the tasks to be tackled has fundamentally changed. In the recent past, policy has been dictated more by ideological concerns about the appropriate roles for the state and the market, but even here a strategic planning framework is still required. This framework is very different from that used in the 1970s and 1980s which was based on the production of large-scale proposals to meet expected shortfalls in capacity. The planning process has become more holistic with the transport elements being linked to housing allocations and the need to maintain regional competitiveness. The process is also more broad based, involving a wider range of stakeholders and affected parties. It has been democratized and become more normative, principally through the introduction of complex objectives such as those related to sustainable development. This contrasts with the more traditional positive approach based on the 'scientific' method and the belief that through careful quantitative analysis one could understand the complexity of cities and evaluate a range of alternative strategies to meet expected levels of traffic demand. Transport planning has recognized the limitations of the quantitative approach and accepted the fact that decisions in transport are essentially political, requiring a wider range of both quantitative and qualitative analysis. In addition, there is a maturity in approach which acknowledges the necessity both to explain what is happening in the methods used and to recognize the limitations of these methods. These issues have been extensively covered in the retrospective part of this book.

Whether investment is funded by the public or private sectors (or jointly), some stability is required in the decision-making framework so that decisions can be made with some certainty, otherwise only low risk strategies will be adopted. This may result in only those transport projects which are underwritten by the government actually proceeding. If the private sector is to be fully involved in the new generation of investment, then a planning framework is required so that the risk and return levels on investments are reasonably well known.

Predicted increases in travel demand cannot be met. The expectations that new capacity could be provided, either through new construction or through traffic management, to meet that growth has now been firmly rejected. Even if it were possible, it would be undesirable. The land required for new infrastructure is at a premium in many countries, and the environmental and social costs involved make it unacceptable, particularly if a solution is only temporary. In many transport systems working close to capacity, additional increases in that capacity will be immediately taken up by 'latent' demand with previous or even worse levels of congestion being quickly re-established (SACTRA, 1999).

The new realism in transport planning recognizes that there is no possibility of increasing road supply to a level which approaches the forecast increases in traffic. Whatever road construction policy is followed, the amount of traffic per unit of road will increase, not reduce (Table 8.1). In effect, all available road construction policies result in congestion getting worse, either in its intensity or in its spread. The only way forward here is to make demand management the central feature of all transport strategies, independent of the ideological or political stance. Transport planning has always been an intensely political activity, but there does now seem to be some agreement on what needs to be done both politically and professionally. This means that radical action could take place provided that public support could be obtained, and that the necessary skills and organizational framework were in place so that the traditional mould of transport planning can be broken.

A third element in the new picture is a vision about the future of the city in terms of the quality of life and its 'liveability'. In the past, the central concern has been over increasing the quantity of travel, the acquisition of the car, and the notion of the freedom to use that car. As affluence increases, other factors related to the quality of travel, the quality of life and environmental responsibility become important – values change. Important economic concerns, such as those relating to employment and local economic competitiveness, are still central to policy-making, but other factors relating to the environment and social justice have also become key elements in the search to promote the sustainable city (Williams *et al.*, 2000).

The fairly narrow conceptual framework used by many countries (the Anglo Saxon approach) allocates a market role to transport with efficiency and productivity being the prime policy objectives. Intervention only takes place where the market is seen to fail or where there need to be adjustments for social reasons. A more continental approach (the French approach) to transport planning assigns transport the status of an intermediate activity that requires direct control to achieve wider social, industrial, regional and national objectives. The state still has a central role to play, but policy is driven much more by national objectives (usually normative) that can embrace a wider range of interests, including quality of life.

The new challenge

Having made a strong argument for new approaches to transport planning and one that places it firmly in the context of economic, environmental and social policy objectives of sustainable development, the next step is to determine what the main challenges will be in the new century. The underlying rationale for transport planning has not changed, as its primary objectives have always been to facilitate access to and participation in activities, but the means to achieve these objectives have radically changed. Moreover, these traditional roles have been extended to ensure that all groups within society benefit (that is, avoiding exclusivity), that broader objectives are met (for example, on the environment), and that the quality of life for all is maintained and enhanced (for example, liveability of cities). Some of these objectives are likely to be in conflict with each other, and in these cases planners should assess all the evidence before advising decision-makers what to do.

One clear lesson from the 'market experiment' in the 1980s has been that the market works well in certain situations, but there is still a need for a clear overall strategic framework within which the market can operate. It is in the creation of this framework that planning has a new role to play. It is really helping to define the social market and the means by which planning can work alongside it. In the past transport planning has tended to limit itself to a narrow conceptual base, looking at problems from the transport perspective and presenting solutions only in terms of transport options. Even within transport itself, the range of options considered has been mainly restricted to pricing and physical measures. Increasingly, problems, analysis and solutions must be examined holistically as all these questions are part of the same process.

At the strategic scale, there is the market-state relationship which has now shifted substantially towards the market. This shift is likely to remain, but the role of the state has to be redefined as supporting the private sector. New forms of transport planning have evolved with regulatory organizations ensuring that the newly privatized transport operators are promoting the public interest as well as their own profit and shareholder related interests. The role of the state is not only one of control, but also one of ensuring fair competition and value for money, particularly where public investment is still involved.

One objective must be to place environmental issues at the centre of planning, which means that economic objectives need to be matched by clear environmental improvement to meet global and local targets. There are major opportunities here for creating solutions that improve both the environment and economic performance. For planning, the role is somewhat more difficult as moves towards sustainable development strategies may result in economic benefits being moderated. For example, in transport there should be clear environmental targets with charges being related to the full environmental and social costs of each journey made. Environmental audits and best practice guidelines are now part of the responsibilities placed on companies to promote the minimization of total resource consumption, to encourage recycling, and to reduce the levels of environmental damage (Maddison *et al.*, 1996).

Underlying much of the debate at the strategic scale is the public acceptability of the proposals. To have a real effect on the behaviour of firms and individuals, there is the considerable task of convincing them that change is necessary – commitment often falls short of real action. No matter how attractive public transport is, no matter how expensive petrol is, no matter how close facilities are located to the home, people will still use their cars. Policy levers such as pricing and control may only have a limited effect on actual behaviour. To hold any expectations that reality is different misinterprets the dependence of current lifestyles on the car and the perceived freedom it provides. One real challenge to the transport planner is to argue convincingly for cars not to be used and for people to accept that argument and to leave their cars at home. It is widely recognized that the acquisition of the car is the most important single factor in changing travel patterns – from the destinations used, to when the trip is taken, to the trip lengths, and to the forms of transport used. The new issue here is that where car ownership levels are universally high, how can transport planners actually influence travel patterns (Banister and Marshall, 2000).

One interesting possibility is the notion of traffic degeneration (Cairns, Hass-Klau and Goodwin, 1998). As discussed earlier, it has been suggested that additional capacity is likely to increase demand. The counter argument is that if capacity is reduced, some travel will be lost. If the use of the car is restricted in city centres (for example, by traffic calming) or through land-use and development policies (for example, by concentrating on city centre development), or through the use of technology (for example, by telecommuting) will travel be reduced? Similarly, in the past the concern has focused primarily on the amount of travel, but now quality of travel is also important. So if that quality declines, will people reassess the necessity to make a journey? Underlying these key questions is an understanding of both public acceptability and receptiveness to radical change, and of their behavioural response.

Radical alternatives are required. Transport systems management was used in the 1970s to increase the capacity of the road network through low-cost schemes such as area traffic control, restrictions on parking, and extensive one-way systems. This was followed by demand management in the 1980s to promote high occupancy vehicles, new public transport systems, parking controls and pricing, and extensive pedestrianization and calming schemes. In the 1990s, there have been more demand management schemes (for example, park-and-ride, bicycle priority, central area management, access restrictions) and a reliance on technological solutions (for example, route guidance, parking guidance). But there is an institutional reluctance to implement a road pricing policy or to use the planning system to limit the growth of traffic at source (for example, through clear strategies on sustainable development).

As a result of this inertia at the strategic level, most of the action has taken place at the city and local levels where responsibilities are clear. It is also at this level that many of the real effects of congestion, inefficiency and poor environmental quality are actually felt. Greater car dependence and higher levels of mobility result in increased congestion as the capacity of the transport system fails to respond. Cities become less attractive places in which to live as decentralization takes people and jobs to peripheral car-accessible locations. To reverse these trends means that cities must become attractive locations for investment, with affordable housing and high quality facilities and amenities. They must be seen as safe, secure and pleasant places to live. Transport plays a pivotal role in achieving such a city, and although the car may still be essential, it must be seen as promoting cities rather than as one of the main reasons why cities have become hostile environments (Tolley, 1997).

What then are the options for transport planners in cities? The car is an inefficient user of road space and this must be realized and accepted by all through charging for parking and for using road space. Smaller city vehicles (including those using clean fuels) offer some potential, but this does not solve the problem of space. Parking controls are the main means available to limit the use of the car in urban areas. The bus (or tram) is the most efficient user of city road space, yet sharing that space with the car reduces efficiency. The urban road network could be designated for particular uses. For example, in the city centre 30% of roads could be allocated to pedestrians and cyclists, a further 30% to public transport and access only, and the remaining 40% to general use. The proportions would not be uniform across the city, but would relate to the dominant land user. The implementation of such a scheme could take place immediately, and the capacity and quality of the bus system would be dramatically increased as scheduled operating speeds would be improved. There would be fewer cars in the city centre and the environmental benefits would also be significant. Segregated bus networks would then become a reality. These proportions could also be made variable to reflect changing priorities over time of day or day of week. The technology is available for the flexible use of road space, in terms of allocation to priority use and in terms of pricing.

Traditionally, the cases for both pedestrian areas and traffic calming have been argued on safety and (more recently) environmental criteria. More extensive schemes could be established in residential areas and quality neighbourhoods could be created where there is no access for polluting vehicles. All travel is by soft modes (for example, walk and cycle) or through electric delivery vehicles or clean public transport. In each case schemes should be seen as being a part of a more general area-wide strategy as the concern is to reduce traffic overall (degeneration) rather than to divert traffic elsewhere.

Apart from actions on individual forms of transport, there needs to be coordination between all forms of transport. Included here are parkway stations and park-and-ride facilities at peripheral sites or at suitable interchange points, where car users can switch to regional and suburban rail, tram or bus services. Car sharing can be encouraged through high occupancy vehicle lanes, and all employers should develop company transport plans to reduce car dependence (for example, for firms, schools, universities, hospitals, local authorities etc). Investment is still required to upgrade and expand the existing public transport infrastructure. There is no lack of ideas here and many cities have addressed these problems, some with very innovative solutions. Similarly, a new optimism can be seen in investments on the high-speed rail networks (Whitelegg, 1993). But in many situations, there is an absence of coordination between the different schemes and in the financing of all schemes. It has proved almost impossible to get the private sector to invest substantial capital in transport infrastructure, even in partnership with the public sector. In addition, the availability of traditional forms of public sector finance have been limited and are unlikely to be renewed, at least to the same levels as before. New financing mechanisms are required in the form of low interest government backed loans (as in France) or through public bonds (as in New York), or through creative forms of partnership finance with the public and private sectors. This again is a real challenge to transport planning.

Into the new millennium

In the longer term, the most important contribution from the transport planner is in a real integration of land-use and development decisions, together with an assessment of their transport impacts. Decisions now being made on the allocation of new housing to meet the demands of smaller households, single person households, and an ageing population will determine levels of traffic generation in the future. Similarly, the location of workplaces, shops, services and facilities will all influence the journey lengths and forms of transport used in the future. This integration will actually tackle the transport problems at its source, namely where the travel is generated, and it has a substantial contribution to make to the sustainable development agenda.

Location strategy, together with clear analysis of density, settlement size, facilities and services, and mixed use development, would all enhance accessibility and provide a choice for individuals not to use their cars. The solutions for transport planning in cities, particularly city centres, are clear, through the creation of high quality environments at intermediate densities with efficient and attractive public transport – there is no reason to own or use the car. But the solutions in the suburbs, rural areas, on the interurban networks and the growth in international travel are much harder. In each case (except the last) the role of the car is crucially important and some people seem to be prepared to pay a high price for transport so that they can live in low-density locations (Williams *et al.*, 2000).

Transport planning must take the lead in understanding and analysing these problems, and placing them in the context of environmental and social concerns. Planning should seek to reduce the environmental impact of transport, both at source (through technological innovation) and on the road (through pricing and regulation). Equally important is the necessity to provide access to jobs, services and facilities for all people, particularly those on lower incomes without access to a car. But accessibility is not just a matter of transport. Accessibility is having local facilities, a strong social network, the means to communicate with others, as well as the means to get there and the necessary resources. Closure of local facilities often results in savings to the producer, but the costs are passed onto the user as they have to travel further (by car) to get to these centralized and specialized facilities. Transport planning must balance the requirements of the market with those of the user, particularly those who are likely to have problems of accessibility. There is a substantial potential here for the new technology to break down many of these barriers. But in reality, more barriers may be raised, as those with low levels of transport accessibility are also those with lowest (potential) access to the new technology. This may provide the greatest new challenge to transport planners, namely to establish the means by which technology can actually replace some car-based travel, but at the same time ensure high levels of access by all to the new technology.

For example, in rural areas where the quality of public transport is limited, there are substantial opportunities for transport innovation through community transport, social car schemes and flexible shared taxis. Technology now provides additional flexibility in the scheduling of vehicles and in being able to respond to requests almost instantaneously. There is a substantial potential synergy between transport and technology in rural areas that needs to be exploited. Even though local facilities have been closed and many forms of rural services are under threat, it may be the new technology that comes to the rescue. Goods can now be ordered over the Internet, but one of the main problems is the distribution of those goods to people's homes, particularly when people are out at work. The local shop or post office could become the local Internet distribution centre where people could collect their goods when it is convenient. For the supplier, it offers the opportunity for less distribution and a guaranteed point of delivery. This is a win-win situation and may allow uneconomic rural services to have new life.

As we move into the new millennium, how will these three great changes continue to evolve? There will also be a reinterpretation of priorities as little new infrastructure will be built, due to the political unpopularity of roads and the need to use available resources to replace much of the existing infrastructure. Most investment will be in the public transport infrastructure and in the means by which technology can be used to maximum effectiveness in squeezing more capacity out of the available transport system. Transport planning will be transformed and take a leading role in providing a link between the different public and private sector organizations so that the full implications of decisions taken in any particular sector can be interpreted. There must be a balance between the narrow market driven economy, and the broader values assigned to environmental and social priorities. There is a huge opportunity for transport planning to become a key player in the sustainable development debate, as transport continues to be a major user of non-renewable resources and a polluter of the environment. Through the use of technology and planning controls, cities can be made clean, desirable and liveable. There is also scope for new methods of environmental evaluation and auditing to become central in all planning, together with new responsibilities for monitoring and achievement of challenging environmental targets – this is the vision of the sustainable city.

Transport needs basically depend on where people live and work. But more importantly, with the new patterns of household structures, with the ageing population, with the growth in leisure based activities, and with the known and unknown effects of new technology, this will all change. Added complexity arises out of the interaction effects of how quickly or conveniently people can travel even over short distances, as this influences where and how they work and live, and this in turn is affected by the other factors mentioned above. If people change how and where they work and live, and what they do in their leisure time, different transport needs will arise. This interdependence continues to become more complex with time, providing a range of exciting new challenges to transport planning.

Implications for analysis methods

These challenges for transport planning mean that new thinking is also required in terms of methods and approaches to analysis. First, the patterns of change need to be reviewed so that the main driving forces can be identified. It is here that much analysis becomes stuck as the trend based 'business as usual' future does not seem to be attractive or able to encompass the scale and nature of new travel demand patterns. Intervention and change is required, but do we have the methods to analyse such change? Secondly, the key concept of uncertainty needs to be explored as this underlies all futures, and it seems that for transport analysis uncertainty is particularly important. Uncertainty must be accepted (and encouraged), but understanding improves if we place transport within a wider context of development through identifying the characteristics of the desirable city of the future. Thirdly, we need to review the methods available for investigating the driving forces, the uncertainty and transport within the context of the city.

Driving forces

The recent Transport 2010 document (DETR, 2000b) does not make good reading. It suggests that road traffic will increase by 22%, and that congestion will get worse by 15% (2000–2010), but worse by 28% on the trunk road network. Overcrowding and congestion will be endemic in London with 40% overcrowding on commuting by rail and road congestion in the capital being 3½ times the England average. Public transport is inadequate and deregulation has not resolved the problem, except perhaps in London. Rail demand is expected to increase by a further 34% (2000–2010), but capacity constraints will limit this increase to 23%. Passenger numbers through airports will increase by 50% to 2010 (see discussion on this in Chapter 5 and Table 5.9).

Two comments should be made here. The focus of the policy statement is very much on reducing levels of congestion on all the transport networks with the conventional argument that travel time is valuable. Perhaps analysis should be moving away from travel time minimization to recognizing that not all travel is a negatively valued activity (for example, leisure time, which in its various forms now accounts for about 20% of all travel and is increasing), and we should be looking for thresholds of acceptable travel time. But the new policy focus has also pushed the environment down the agenda. There is nothing in the Transport 2010 document on meeting CO_2 reduction targets in the transport sector, but other forms of local air pollution will be reduced (at least to 2010), through innovation in vehicle technology and improvements in fuel efficiency and quality.

Nevertheless, there is a strong expectation that demand for travel will increase. There seems to be little evidence that the assumed link between economic growth and transport growth has been broken, and the transport intensity of the economy is still at historically high levels (SACTRA, 1999 and Banister and Berechman, 2000). It is important to get behind the figures and understand the driving forces.

Demographic change is having a fundamental effect on travel demand, resulting from the ageing population, changes in household and family structures, delays in starting a family, ownership patterns with respect to housing and cars, and the suburbanization process. Even though 60% of the poorest households have no car and 55% of those over 70 have no car, this will change (DETR, 2000*b*). The aspirations of most people seem to be to join the car owners, and the elderly will keep their cars as long as possible. Many drivers have now had experience of cars all their adult life. For example, those born in the 1940s, acquired a car in the 1960s and will keep it until 2020. The demographic impacts have been recently strengthened by the new flexible work patterns (both in temporal and locational terms), by new patterns of location and distribution (for services and internet shopping, for example), and by the growth in non-routine travel. Work related activities (work and business) account for under 20% of all activities, but even these are now less regularized or routine. The growth areas for travel are in the other 80% of activities, many of which are spontaneous activities. So there is a new dynamic in travel. Even the pattern of 'regular' trips is becoming less regular, less important, longer distance, and more car dependent. The growth market is in new 'spontaneous' trips arranged by the new mobile technology whilst in transit. This is the challenge for analysis.

Organizational change is also having a fundamental effect on travel demand. Business and industry have been following a trend of globalization with larger companies acquiring smaller ones, supplemented by strategic mergers and alliances. The underlying argument here is that there are substantial economies of scale and profits to be made. For example, in the motor industry, it is likely that there will be five or six mega producers worldwide by 2010. But there are other changes taking place in industrial networks as companies increasingly outsource their production, with suppliers having to produce quality and quantity to tight schedules. Such change has been facilitated by technology, but it has resulted in many suppliers locating near to their main partners to reduce the unreliability inherent in the transport system. Risk is reduced in this respect through location strategies. More generally, it is still argued that there are substantial agglomeration economies brought about through concentration (Anas *et al.*, 1998 and McCann, 1995).

Underlying the globalization process is the new 24 hour economy with flexible working, and a clear distinction between core workers and others on different contracts – what Handy (1995) calls the Shamrock organization. Again here we have a new dynamic in travel as producers require complete reliability in their supply chains and as production processes become compressed. Increasingly, production will be market led. This is already happening in the computer business where products are not bought off the shelf, but are ordered by the customer and then made to their specification and delivered in a week or less. The same will happen with car production, although the delivery guarantee may be slightly longer at about two weeks. The service and information based economy is driven by customer requirements.

Technological change is also having a fundamental impact on all travel as we are currently in a revolution that is likely to be just as important as the agricultural or industrial revolutions. Technology impacts on all activities, whether it is to be found in vehicles, in transport systems, in communications, in information, in gaining access to services and facilities, in work etc. The difficulty about exploring technological futures relates to questions such as diffusion of technology, the rate of development and use of online services and facilities, and the role of the e-economy. Some of these issues have been discussed elsewhere (Banister *et al.*, 2000), but it is important here to be aware of the potential and limitations (if there are any) of the new technological revolution. Certainly, in a more pragmatic sense, the availability of personalized information and communications systems with secure payment devices will begin to make multi-modal travel more attractive to individuals, and to provide the necessary links between all modes and locations. The need to own a car in the city should also be substantially reduced.

Uncertainty

Even though political and economic stability are taken for granted, there is increasing uncertainty about the future, and analysis should recognize this. At all levels of decision-making, there is greater variability in both the temporal and the spatial characteristics of travel. By definition, analysis and modelling attempt to simplify complex processes, but the balance must be sought between what is useful and what is possible. Two potentially useful approaches to the handling of uncertainty as a context within which analysis takes place can be identified.

First, it is useful to identify contextual (or external) factors that are likely to remain stable over time or outside the process, at least over the next 10 to 20 years. Included here are the stable political and economic environment within which transport policy operates, as for example reflected in the expectation that economic growth will continue. Some of the driving forces of change are also external factors (that is, demographic change, organizational change, and technological change). Even within the transport sector, other assumptions can be made, some of which might be more debatable as they do not fall exclusively outside the transport analysis arena. Included here are the political objectives of maintaining the market orientation with the user paying the full costs of the transport services used, the continued operation of most transport in the private sector, and the rapid take up of the latest logistics and technological systems in transport. It is important to be able to establish at an early stage in analysis what is assumed to be external to the process, particularly in light of the uncertainty about the future. In much analysis, these factors are assumed, but not stated.

The second complementary means to encompass uncertainty is to think about desirable futures in terms of a visioning exercise. This is where transport must be seen as part of a much wider process. What are the characteristics of the desirable city of the future in terms of its employment base, its functions (for example, as a leisure resource), its quality of life, its safety and security, its development strategies, its policy on affordable housing, its culture, its diversity, its environmental quality, etc? Policy should be directed towards improving those key characteristics of the desirable city, and transport has a major role to play in that process. The argument here is that thinking must move beyond the short-term responses to perceived problems, to establish a much clearer long-term vision as to how urban development can and should develop over the next 20 years (Hall and Pfeiffer, 2000).

Too often, researchers only feel comfortable when faced with detailed problems of a specific nature to resolve. They feel uncomfortable when faced with the much broader questions of uncertainty (in a general sense) and participating in the debates about desirable city futures or the impact of technological innovation. There is a need for all researchers to engage in the more strategic debates and take a broader view of the impact of local solutions on the network as a whole, or in terms of achieving stated policy objectives in transport and in public policy more generally.

Methods for looking into the future

In essence, there seem to be three complementary approaches that can be adopted to analysing longer term futures in transport, all of which have their own advantages and disadvantages (Van Geenhuizen *et al.*, 1998).

Modelling approaches – These methods have provided the main means by which futures have been explored in the transport sector, either through extrapolation of trends or through the testing of different policy options within clearly defined contexts. The underlying assumptions are that patterns of behaviour can be established from empirical relationships in the data, and that it can be expected that these relationships are stable and will hold in the future. There are many well known critiques of these methods ranging from detailed analysis of the methods used (Lee, 1994 and Supernak,1983) to more general reviews of the underlying rationale in the structural and conceptual frameworks developed (Sayer, 1976 and Batty, 1989). The main issues here have already been extensively covered in the retrospective part of this book (Chapters 1–6).

However, transport modelling has remained a remarkably robust science with its basic underlying structure and rationale remaining intact even after attack. It has reacted through consolidation and defensive action as well as taking the offensive through the development of new initiatives. Some models have become more focused, looking at particular elements in the process, but others have explored the implications for transport of different development strategies through GIS based analysis. These approaches seem to be quite effective at identifying change over relatively short periods of time, given stability in the assumptions used, particularly on behaviour. However, they are less appropriate for longer term predictions, or for situations where there is a high degree of uncertainty, or where trends may end, or alternative futures are seen to be desirable (Ayres, 1998).

One interesting corollary to the modelling approaches over the longer term is the presence of cycles and discontinuities in the processes. Much has been made of predictable cycles in the physical sciences (for example, eclipses or the pendulum movement of a clock), but it is only the seminal work of Kondratieff that is commonly cited in the social science literature (Kuznets, 1930 and Schumpeter, 1934). These cycles relate to the acceleration of economic activity and the rise in commodity prices following an innovation in the bottom of the cycle (for example, power technologies). This gives rise to the upward cycle which peaks and decline takes place in the downward cycle over a period of about 50 years. The question here is whether and why the cycle needs to be of a constant length. Perhaps with the current technological innovations the cycle is reduced to being imperceptible, or it may be part of a much longer cycle – hence the 10 years of continuous growth (unprecedented) in the US economy.

Discontinuities result from external unpredictable events, either of a natural cause (for example, earthquakes) or of anthropogenic origins (for example, war). Such events are usually excluded from predictions about the future as change is assumed to be gradual and smooth. But if a paradigm shift or a major change in policy direction is to take place, then this might well result in such a discontinuity. This argument may be particularly relevant in the debate over sustainable transport where the emerging view is that current patterns of consumption are unsustainable, but that all policy actions are not sufficiently strong to change direction in a fundamental way. A discontinuity of substantial proportions (for example, a health epidemic that is transport induced) may be the only way to achieve major change, but even here there are doubts over whether real change will actually take place. In the longer term, the same well established behavioural patterns will be re-established (Banister, 1997b).

Participatory approaches – These approaches are being used increasingly to achieve policy change as it is now realized that involvement of all actors in the process is essential to get their commitment to real change. In the context of sustainable transport, it is the argument that all people and companies are part of the problem and so they should all be involved in seeking appropriate solutions. It is not just a matter of policy intervention (for example, through regulation and pricing) and actions in the public sector, but an acceptance by all of the need to reduce resource consumption and to adopt a higher level of social, moral and ecological responsibility. Without that acceptance for the need to change behaviour, the effectiveness of policy intervention is reduced and may even be counterproductive. For example, to avoid paying a cordon toll to enter the central area of the city, shoppers may drive to a more distant alternative centre where there are no charges – or firms may seek to relocate outside the cordon area.

Participatory methods attempt to (Van Geenhuizen *et al.*, 1998):

- collect relevant information from all parties;
- reveal stakeholders' values, perceptions, expectations and options;
- establish a better relationship and communication between stakeholders;
- enhance frame-reflecting learning, particularly to understand the perspective of other stakeholders;
- enhance and facilitate creativity in developing and enhancing solutions or in creating new ones.

These types of approaches are commonly used in the Netherlands and in the UK to involve all parties in the debates through setting up panels or citizen groups so that a creative process of discussion, information and conflict resolution can take place. One such method used is *creative steering* where a limited form of chaos is introduced to realize innovative decision-making. Alternatively, looser methods of *discourse analysis* can be used where the animator mainly listens to the views of the group, but at certain points in the discussion tries to guide the process or to get some consensus views (Healey, 1997; Vigar, 2001). In both cases, the purpose is to encourage participants to articulate their views on a particular issue, to listen to other views and to come to some resolution. However, as with all such creative processes, expectations may be raised and the outcomes are often different to the intentions, both at the individual level and more generally. This is because the real powers and responsibilities are not reflected in the participatory group dynamic.

There seems to be a clear set of key factors that influence the effectiveness of participatory processes in looking at the longer term futures for transport (Van Geenhuizen *et al.*, 1998, p. 207):

Motivation – all relevant actors need to be sufficiently motivated to participate. They need to be convinced of the problem and convinced of cooperation as an important way to arrive at a solution.

Transparency – of aims and procedures for the participation of all stakeholders and trust. Stakeholders need to be convinced of a potentially genuine participation, as opposed to symbolic and manipulative participation.

Barriers between stakeholders – barriers from different 'languages' and types of argumentation need to be prevented; the match between the type of approach and the 'context' of stakeholders deserves a strong concern.

Role of the process manager depends on the aim of the participation – the role is critical in terms of the interactions between various stakeholders as a facilitator, moderator, adviser, or collaborator.

Time between the participation and implementation – the positive change in attitude of participants cannot be sustained over a long time period (e.g. a maximum of two years). So implementation is important as a demonstration effect.

Participatory methods are important as a means to obtain a range of views and the involvement of all stakeholders. However, their use is limited to the shorter term future and to the means by which alternative strategies can be discussed in terms of actual implementation. As such, they are an important adjunct to the traditional modelling methods where they can be used as a front end module to assess different options to be tested in the modelling process, or as an add-on module when implementation of alternatives are being debated. When participatory methods are used to look at the longer term future (that is over 10 years), their usefulness is more limited as they have less experience of thinking about the types of change needed to move towards a more sustainable transport system. Their role is still important, but preliminary research has to be carried out to specify particular types of innovation or change which can then be discussed. It is necessary to focus on particular issues or questions rather than to have a very general debate about the future.

Scenario analysis – Tremendous interest has been shown in the use of scenarios for transport analysis. The CEC defines a scenario as:

> a tool that describes pictures of the future world within a specific framework and under specified assumptions. The scenario approach includes the description of at least two or more scenarios designed to compare and examine alternative futures. (CEC, 1993)

There is an acceptance in scenarios that the future is uncertain and that we can no longer rely upon past trends to give us an insight into alternative futures. The rate and scale of change is phenomenal, both in terms of societal and technological change. Scenario analysis is increasingly being used in long-range policy research, since it provides a way of identifying future issues and problems for policy-making in an environment of qualitative uncertainty. The scenarios develop alternative futures in a structured way, which in turn helps to identify policies that are robust across a wide range of possible futures – they are not concerned with any one particular future. When examining trend-breaks, for example through the introduction of new technology or through policy initiatives (for example, on sustainable development), it is important to explore futures according to a different logic that can accommodate uncertainty through its focus on causal processes and key decision points.

Scenario analysis has an important role to play if the following conditions are met (Nijkamp *et al.*, 1998):

◆ The number of stakeholders and relevant factors is high.
◆ There is controversy between stakeholders about the actions to be undertaken.
◆ Uncertainty about the future is large, including new events but the system as a whole is stable.
◆ Policy-making is relatively open.

The new agenda of a sustainable transport system for the future meets all four of these conditions.

A scenario describes the present situation in society together with likely and desirable future states of society, and a sequence of events that can link the present situation and future states (Becker, 1997). Becker also distinguishes between two basic types of scenario approaches. In the *American approach*, a distinction is made between context and strategy with the scenarios first being presented as the context within which the system operates and policy making takes place. Various actors are then asked to choose between alternative strategies and to adopt these so that 'least regret' strategies can be selected by the user of the scenarios. It is essentially a positivist forecasting process where the trends are modified by the inputs from the actors. In the *French approach*, a comprehensive picture of the future is presented in terms of the current situation, and this is supplemented by a description of some future alternatives and of a number of events which may connect the present situation with the future ones. This process is normative with desirable futures being identified (Rienstra, 1998 and Banister *et al.*, 2000).

To Becker's (1997) list, the *Swedish approach* (an adaptation of the French Approach) can be added. The distinguishing characteristics of the Swedish approach is that they are explicitly normative and that they use a backcasting approach where an image of the future is constructed that takes account of current trends, but also implicitly accepts uncertainty in the future (Dreborg, 1996). A path is then constructed on how to move from where one is at present towards this desirable future position. The scenario targets, the images and the policy paths are all validated at various stages by experts, so that feedback can take place to modify the scenarios. The process is not prescriptive, but illustrative of possible future paths and gives an indication of the nature and scale of actions (together with their sequencing) of the changes necessary to achieve the scenario targets.

There is a role for scenarios in both conventional policy analysis and in evolving trend breaking alternatives. In the former, the main concern is to help design policy packages that meet clearly defined policy objectives. One strength of scenario analysis is in investigating how policy measures can work together to achieve targets or other objectives. In the latter this concern is extended to the debate concerning desirable futures, such as those based on sustainable development, so there is also a clear normative element. Both Becker (1997) and Nijkamp *et al.* (1998) have argued that scenario analysis has the potential to promote learning and a better understanding of complex situations by policy-makers and other stakeholders through:

- the stimulation of creative thinking and communication – conventional alternatives are challenged;
- the identification of key events or forces which are relevant to policy resolution or the prevention of undesirable situations – critical incidents;
- the simulation of alternative policy options together with an assessment of the necessary preconditions and the possible outcomes;
- the explicit inclusion of uncertainty within policy analysis so that flexibility and robustness can be included.

All three approaches to exploring the future are important and they are not mutually exclusive, but mutually supportive. They should be used in conjunction with the more conventional approaches outlined in the retrospective part of the book. However, it may be time now for a more radical approach to policy analysis and forecasting. Approaches should be trend breaking rather than trend following. The time for incremental change in well established methods may now be past, as futures based on the concepts of sustainable development must embrace trend breaking alternatives and accept the important effects of the driving forces and the notions of uncertainty outlined here.

A new perspective on the transport agenda has been presented that focuses on the main changes in policy and planning, and attempts to look into the shorter and longer term futures. However, the future itself is becoming increasingly unpredictable, but this does not mean that this challenge should be ignored. It suggests that new approaches should be developed with even more creativity so that both trend following and trend breaking futures can be explored. All analysis processes must involve more participation and discussion with all concerned stakeholders. The era of the expert is long past and the only means to anticipate future change and to introduce real policy change is through inclusive participatory analysis, as all parties must understand the need for change and be prepared to contribute to that change.

Note

1   Predict and provide is the notion that the provision of transport infrastructure is demand led and that new capacity (mainly roads) would be constructed to meet that demand. The reality suggests that this has never occurred as there has never been a period when sufficient investment resources have been available. Demand has always grown at a faster rate than capacity.

Chapter 9

The role of transport planning

Introduction

Transport planning has now come full circle. Since 1960, its role has developed from analysis based on open-ended inductive experience to large-scale systematic analysis involving the systems approach to transport planning. It was only at the end of the 1970s that the most significant changes took place, with the initial broadening of concerns over the distributional effects of decisions, with the environmental concerns and the externalities of transport, and with the difficult decisions resulting from the restructuring of the economy. These policy concerns meant that transport analysis, at last, was seen as a central political issue and not one that could be left to the expert. In the 1980s, the planning framework was dismantled as the principles of the free market economy were imposed, as transport enterprises were returned to the private sector, and as public expenditure was cut back. In the 1990s, new challenges emerged as described in the previous chapter, and there has been a reversal back to the plan led approach with the social market playing an increasing role. If these challenges can be met, then real progress will have been made in terms of addressing congestion, determining the role of the state, and in resolving the environmental issues.

The lessons from the recent past might suggest that there was no role for transport planning. This conclusion would be mistaken. As described in Chapters 5, 6 and 8, there has been a renaissance in terms of the techniques and approaches used, and in the problems being addressed. Closer links are being established between land use and transport, new behavioural and dynamic techniques have been developed, the use of geographical information systems and graphical user interfaces is now becoming common, and a wide range of databases are being extensively developed. The key problems of demographic change and industrial restructuring are being supplemented by demands for new infrastructure, and the potential for the new technology in transport is now being comprehensively tested. These changes represent the European and national agendas. At the regional and local levels there are also key transport issues to be addressed, particularly where the market cannot work but a service needs to be provided. It is here that planning needs to operate as an alliance between state and the market. 'The market cannot produce optimal solutions for efficiency, nor socially desirable solutions in terms of equity' (Lichfield, 1992, p. 155). It is this alliance that is now discussed together with the levels at which and the priorities with which transport planning should operate.

The market–state relationship

The experience of the last 20 years has shown that some form of planning is essential; one which is based on alliance, not on conflicts between central and local government or on the dichotomy between the public and private sectors. Various forms of partnership must be identified and the case for such an alliance is clear. New thinking is required to cover at least five important questions so that a social market for transport can operate effectively:

+ There is a need for *intervention in the market* with new forms of regulation. Even the private sector favours a strategic framework within which to operate as stability in political objectives reduces uncertainty and risk, and these two factors in turn allow greater scope for private sector action. Such a strategic view does not conflict with the free market as it would still permit businesses to pursue their own profit-related objectives. The publicly-planned strategic framework would provide the longer term horizon.

+ *Environmental issues* must now be seen as integral to all transport policy decisions and investment proposals. Decisions must be based on a balanced view between value for money and the externalities created. If a serious move towards sustainable development is to be achieved, then economic evaluation has to be moderated by political accountability as there is a considerable cost to becoming sustainable. All forms of transport could be given environmental targets with green taxes or green grants being used to ensure all users are aware of the costs and benefits of using particular modes. Environmental audits could be carried out on firms and schemes as they are implemented, to reduce substantially the total resources used, to increase the levels of recycling, and to limit the damage to the environment. Such an audit could take place on proposals to close facilities as well as decisions on opening new facilities.

+ Underlying much of the debate over the need for radical change is the *public and political acceptability* of any set of proposals. To make any real change in people's travel patterns and to reduce the use of the car, a fundamental switch in attitudes is required. People do not seem to be prepared to make a change in their lifestyles, as any form of commitment falls short of real action. No matter how attractive public transport is, no matter how close facilities are located to the home, no matter how expensive petrol is, people will still use their cars. Policy levers such as pricing and control may have only a limited effect on car drivers' behaviour. To hold any expectations that reality is different is unrealistic until public attitudes and priorities change. There is evidence that this change has taken place to some extent in European cities (for example, Engwicht, 1992 and Banister and Marshall, 2000), but in at least as many other situations such a change is becoming more difficult as more people acquire cars and as lifestyles through choice become more car dependent. Perhaps this task is too difficult, and should not be attempted and the politicians should continue to muddle through rather than take a decisive lead.

• *Traffic degeneration* must form an integral part of any thinking. It is now accepted that as new road capacity leads to a reduction in travel costs, more travel will result, and this in turn will lead to changes in accessibility brought about by changes in land use and the location of activities. New traffic generated as a result of new capacity forms one of the most difficult parts of current road assessment. In most cases, it is ignored and assessment concentrates on the redistribution of existing traffic supplemented by the secular growth trends in traffic. The use of fixed travel matrices is a major limitation in much analysis. Equally important is the understanding of the conditions under which traffic is actually lost, either through restricting the use of the car (for example, by traffic calming), or through land-use and development policies (for example, by concentrating on city centre development), or through the use of technology (for example, by telecommuting). As the quality of travel deteriorates people may reassess the necessity to make the journey.

• *Reducing the need to travel* is one means by which the planner can have a direct impact on travel and the quality of life. Decisions made about the location of new housing, and of services and facilities determines to a great extent the future travel patterns and the modes of transport used. One of the great trends over the last 30 years has been the growth in travel distance which has far outweighed the increase in trip generation. As trip lengths increase, so does dependence on the car and the consequent reductions in the use of green modes of transport and public transport. Reducing the need to travel means making more use of local opportunities, preferably using green modes and public transport. Where the car has to be used, then trips should be organized to reduce total distance travelled (that is, trip chaining) and car occupancy levels should be increased.

Radical alternatives are required. Traffic management was used in the 1970s to increase the capacity of the road transport network through low-cost schemes such as area traffic control, restrictions on parking, and extensive one-way systems. This was followed by demand management in the 1980s to promote car sharing, new public transport systems, parking controls and pricing, and extensive pedestrianization and calming schemes. In the 1990s, most options had been tried, but there were further possibilities brought about by the use of technology (for example, route guidance systems, parking and travel information systems). There must be limits to the effectiveness of traffic management and the ingenuity of transport engineers to squeeze more capacity out of a given transport system. The full potential of demand management (principally road pricing and parking pricing) seems to have been resisted, even though London may experiment with cordon pricing in 2003. The use of planning policy to limit the growth in traffic at source has also not yet been fully recognized.

Underlying much of this reluctance is the question of public acceptability, which seems not to have been tested in Britain. There is an awareness of the problems caused by congestion to the economy through wasted time and resources, to the environment through levels of pollution, and to the health of individuals through increased stress and frustration. Yet no government or local authority seems to be prepared to tackle the problem. Even though the necessary primary legislation has now been passed (the Transport Act 2000), and it is now possible to introduce road pricing, the government has given the cities the responsibility for taking the decision to implement it. This has reduced the attractiveness of road pricing, as local authorities are reluctant to be the first mover, as there is a fear of public reaction and of competitive disadvantage if only one city made such a decision.

In other countries the problem of public acceptability has been tested through a referendum. The state of California and the cities of Zurich and Amsterdam have all used this means to assess whether political action would be supported by the electorate. California has taken the lead in imposing strict environmental limitations on car emissions, in encouraging zero emission vehicles, in using land-use designations to reduce levels of trip generation, and in investment in public transport. Most emphasis has been placed on the journey to work problem and employers are seen as having a clear role to play in reducing trip rates. It is ironic that many of the radical proposals have been tested in the United States where the car culture is most dominant.

In Europe there is greater acceptance of restraint on the use of the car and there are strong planning controls. However, the concept of the compact city, advocated by the European Community (CEC, 1990), is continuously being weakened with peripheral developments, business and science parks, new towns, and the growth in international travel. The traditional radial patterns of movement are disappearing as cities change and as commuting ceases to dominate. Growth has taken place in circumferential movements, in new trip patterns for social and recreational activities, and in longer trip distances, and the net result has been the development of an enormously complex set of travel patterns, which are difficult to undertake by non-car modes of transport.

The funding of transport investment

A constant theme within transport planning over the last 40 years has been the appropriate levels of capital and current funding in the sector. It is generally accepted that under investment in new projects and maintenance has resulted in a system that is often working near to or above capacity for much of the day. The system is also unstable as it only requires a small event to have an enormous impact. The famous butterfly flapping its wings in China and causing chaos in London. Part of the explanation for this situation is that transport has never been given a high priority by successive governments, nor has it had a strategy or vision to which governments might have a strong allegiance. The nearest to this position that policy came was in the radical Thatcher years (Chapter 4) where a strong market based strategy for transport was pursued. But even here, there was a substantial learning process and transport became one of the laboratories for testing new market based ideas. The other main explanation has been the overriding power of the Treasury in all key decisions. Macro-economic policy has determined what resources were available for distribution to the government departments, and it was then left to Ministers to decide how that 'cake' should be divided. Transport always seemed to get a small share. Under investment in the sector has continued from the early 1970s when strict expenditure limits were imposed externally on the UK government.

Two key questions need to be addressed with respect to the funding of transport investment.

1 The role of the public and private sectors individually

In the past, most transport projects have been funded by the public sector as they are large in scale, involve a high risk, and have long payback periods. The private sector has only shown real interest where it has some degree of a monopoly position (for example, in road bridges across estuaries) or where the public sector has underwritten much of the risk in the project (for example, the European high-speed rail system).

However, public budgets have increasingly come under pressure and transport budgets in particular have been easy to cut as their immediate effects are not apparent. It takes many years for under investment in both capital and maintenance to manifest itself in terms of lower standards of service reliability and reductions in safety standards. It is also unrealistic to expect this situation to change, at least in the short term, so there must be a greater involvement of the private sector in partnership with the public sector. For the EU countries, the Maastricht Treaty has defined the limitations for budget deficits and there is now a consensus among member states that the stabilization of pressure on income tax and social security contributions is a key element in maintaining the competitiveness of a single market and in obtaining convergence between member states. The question here is to determine the appropriate role for the public sector as funds will always be in short supply, but is it realistic to expect the private sector to take over the traditional role of the public sector in financing infrastructure investment? The answer is probably no.

The problems of involving the private sector in transport infrastructure projects are well known (Banister *et al.*, 1995 and Gérardin, 1990, 1993, 1995), and they are summarized in Table 9.1. In addition to the central question of risk, transport projects have themselves proved difficult to assess. Levels of demand have been difficult to predict accurately, there are often substantial cost overruns, and they have aroused considerable public opposition to their construction (for example, the TGV Méditerrannée). There is also the important problem of perception as transport infrastructure has been traditionally considered as a free good, particularly at the point of delivery. To compensate for the high levels of risk and uncertainty, the private sector is looking for higher levels of return, typically 12–15%. Very few transport projects fall into this category as most of these high return links have already been built (Gomez Ibanez and Meyer, 1995).

In addition, there are the political uncertainties. For example, the level and legitimacy of toll collection may be jeopardized by inappropriate action from the politicians, as in the case of the concession for the second Tagus crossing

Table 9.1 Summary of the risk elements in transport infrastructure projects as perceived by the private sector

1	Long periods between the start of the investment and financial returns to investors - negative cash flows during both the construction and initial operating phases (5–10 years).
2	Irreversibility of the investment, often involving substantial sunk costs - it is costly to withdraw from a project once it has been started.
3	Financial returns do not flow until the whole infrastructure is completed.
4	Political influences are still manifest in the production of transport goods and services – uncertainty increased through delays from political, legislative or environmental interventions.
5	Long amortization periods in transport projects when loan repayments are required over a much shorter period - the life span of many transport projects range from 20 to over 100 years.

Source: Banister, Gérardin and Viegas (1998)

in Lisbon in the summer of 1994 (Banister, Gérardin and Viegas, 1998). At this level it seems that the odds are stacked against the private sector being fully involved – there is no level playing field. In many countries and cities, the dominant supply is through free access to infrastructure and subsidized operations, and there is protection against innovation or improvement in competing infrastructure or services, except if these are also carried out under concession arrangements.

2 The role of the public and private sectors jointly

Joint projects are a creative concept by which both parties to the contract are recognized as complementary servers to the implicit third party, the public. Both the public and the private sectors have an important role to play in the construction, renewal and maintenance of the road, rail and air infrastructure. Much of the funding will remain the responsibility of the public sector and the planning process needs to be strengthened so that the full range of partnership possibilities can be realized (Table 9.2). This is what can be termed a constructive or expansive partnership between the public and private sectors where complex financial packages and risk sharing determine the exact nature of the agreement. There are also two other roles for the private sector. One has been important in the United States, through the extraction of special real estate taxation based on the added value of land brought about by transport investments (Cervero and Landis, 1995). The other is the creation in France (1982) of the *Versement Transport*, a special tax paid by employers of ten or

Table 9.2 Partnership possibilities between the private and public sectors for transport infrastructure investment

Function	Public sector model	Partnership possibilities			Private sector model
		Type 1	Type 2	Type 3	
Planning	Public	Public	Public	Public	Private
Design	Public	Public	Private	Private	Private
Construction	Public	Private	Private	Private	Private
Operation	Public	Public	Private	Private	Private
Ownership	Public	Public	Public	Private	Private
Finance	Public	Mixed	Private	Private	Private
Risk	Public	Public	Joint	Private	Private

Notes
Examples of the different types
    Type 1 – TGV financing (see Chapter 7)
    Type 2 – Birmingham Northern Relief Road (see Chapter 3)
    Type 3 – Channel Tunnel Rail Link (see below)

more staff and based on their salary volume, which has been used to promote public transport (see Chapter 7). In each of these cases it has been argued that public transport has both direct and indirect benefits to the city, and that these benefits should be capitalized.

The private finance initiative and public private partnerships in the UK

In the United Kingdom, there has been substantial interest for sometime in using private finance in infrastructure projects, but it was only with the Channel Tunnel construction that it became a reality. In 1992, the Private Finance Initiative (PFI) was announced as a specific commitment to the development of a more efficient transport system through the introduction of private sector management expertise into projects. It was also seen as an opportunity for government to transfer risk to the private sector. The particular concern was that public projects had weak risk management, which resulted in substantial cost overruns. However, the timing was unfortunate as it coincided with the decline of capital expenditure by government in transport (see Tables 4.2 and 5.7). This meant that the private sector was in a particularly strong bargaining position. But the PFI could itself have contributed to a downturn in the construction industry as the PFI schemes have not generated sufficient work to compensate for the reductions in the public sector work. There were also concerns from industry over the costs of bidding for contracts, and this led to the exclusion of some small companies. The private sector also found problems with the risk management involved in the larger projects (Table 9.1), and there was still the problem of where the money would come from. The cost of capital for the private sector was often higher than that for the public sector, as the latter was underwritten by government guarantee.

The Channel Tunnel Rail Link (CTRL) provides a good example of the difficulties in negotiating a deal with the private sector (Table 9.3). The review of the National Audit Office (2001) made three key points:

- The restructured deal in 1998 was in many respects more robust than that of 1996, as it made a better allocation of risk.
- The new complex financial requirements agreed require 'significant and long-term' government support for the project.
- The economic case for the link is not clear and the benefits are dependent upon the wider economic and regeneration effects envisaged.

The lessons from these key points are important, as they reflect the difficulties in forecasting demand and returns from the investment, particularly where a substantial part of these will be assumed development benefits at Stratford and Ebbsfleet. They also demonstrated the necessity for

Table 9.3 The progress of the Channel Tunnel Rail Link

	108 km high-speed rail link (300km/h) from the Channel Tunnel portal at Cheriton (Kent) to St Pancras terminal in central London. The project will cost £5.2 billion (2001 prices) and cut 33 minutes off the journey time to Paris (to 2 h 20 min) and Brussels (to 2 h).
February 1996	DETR awarded London and Continental Railways (LCR) the Channel Tunnel Rail Link (CTRL) contract. LCR would finance, build and operate the link through its operation of Eurostar (UK) and money raised on the capital markets. DETR would provide a £2 billion subsidy as the route would also be used by domestic rail services.
January 1998	LCR could not raise the finance and requested an additional £1.2 billion subsidy. The contract was restructured without a 'material increase' in the level of public subsidy. It was now in two sections:

Section 1 – Channel Tunnel to near Ebbsfleet in south-east London;

Section 2 – Ebbsfleet to St Pancras.

Railtrack was brought in to manage the construction of Section 1 and to purchase the line when completed- total costs £1.9 billion (2001 prices). |
| October 1998 | Construction of the Section 1 began with a target completion date of September 2003. |
| July 2001 | Second stage of construction started. To be funded by public and private money – total cost £3.3 billion (2001 prices). Estimated to generate 50,000 new jobs in East London and to contribute £1 billion to regeneration benefits (no source). |

The completion of the entire CTRL scheduled for 2007.

strong government commitment to the project as the risk cannot be passed on completely to the private sector. The restructured agreement clearly addresses this problem with the acceptance of a greater role for government. In the CTRL case, there were also a series of other important constraints. Part of the funding was to have come from the revenues obtained from Eurostar (UK), but the forecasts here have been too optimistic. Although patronage levels have increased substantially, the forecasts have not been met and particular incidents (for example, the fire in the Channel Tunnel) have resulted in uneven results. The private sector should have been aware of the optimistic forecasts and anticipated the financial consequences of lower levels of demand. It seems that in the CTRL case, London and Continental Railway (LCR) shareholders were not at risk, but that it was the taxpayer who was exposed to a financial risk of some £360 million (NAO, 2001). The transfer of Eurostar (UK) to LCR was premature as the planned completion of the balancing external funding for the project was not in place. Normally, the transfer of assets in a Public Private Partnership (PPP) only takes place when all the funding is in place.

Comment

The resolution of the funding crisis in transport is probably the most important challenge over the next 10 years. A start has been made through the Transport 2010 document (DETR, 2000*b*), but as already described (see Chapter 5), even investment of the scale envisaged will not reduce congestion significantly. Only slight changes in some of the assumptions used will result in even more uncertainty in the system. Even if there was the funding available, there are many other barriers to overcome before that investment can take place (for example, the planning and public consultation processes). This means that the interests of the private sector may be limited if there is further uncertainty introduced into the decision-making process. Apart from the complexity of the transport planning process and the time necessary to take a project from its inception to completion, the external constraints on the construction industry must also be accepted. The construction industry is subject to the property market cycle and it is often taken as a 'barometer' for the economy as a whole. Most investment and construction takes place when the economy is on the rise with good prospects for finding a client, low interest rates, buoyant market conditions, and the availability of the necessary skills. When all three elements (transport, financial support, and constraints on the construction industry) are positive, then there is real scope for PPP. Conversely, when any or all of these elements are not working together, the prospects are not so good.

The conclusion reached on funding is that major infrastructure projects must remain a public sector responsibility for the planning processes and for funding. The private sector has an important role in delivering the service to the user through the operation of the system. A mechanism has now been found to take capital expenditure on public infrastructure out of the government's annual expenditure plans, as expenditure is now seen as investment.[1] Such an approach has been developed and used in France to finance major infrastructure projects, where loans from commercial banks and the European Investment Bank are used, but always with government guarantees. This means that the interest rates charged are lower (typically 1–3% less) than commercial rates, and the rates of return on investments can be lower (as they typically are in transport). The net effect is that projects can be funded more cheaply than if the private sector is involved, and those costs can be repaid over a longer period. As described in Chapter 8, SNCF and RFF can both run at a loss and pay no tax to government, at least in the short term.[2] This may result in profits being made on the investment over a much shorter period than if private finance has been used as the interest payments are much lower. In the meantime the user enjoys a railways system of a high capacity and high quality.

The planning framework

Planning must operate at all scales and provide the framework within which the market can operate. It must modify the market where there is a necessity to balance efficiency against other objectives. This role would be complementary to the political objectives set at the national and international levels which would ensure that fair competition takes place, that quality standards are observed, that regulations are set, and that barriers to movement are eliminated. The responsibilities of planners extend beyond the narrow interpretation of planning in terms of physical and economic planning. These responsibilities would include issues such as planning for institutions and finance, technology, energy, the environment and planning for people. These themes should cut through the basic grouping of factors identified in Table 9.4 and they have impacts at all levels.

Within this framework, five main issues are covered, where it is seen that transport planning has a major role to play in determining and realizing change:

◆   economic growth;
◆   promotion of economic development;
◆   environment and health;
◆   cities for people;
◆   meeting social needs.

These issues relate to the middle column in Table 9.4 and form the central

Table 9.4 The basic physical and economic components of the planning system

Scale	Issues	Methods
European Union ↕ National	Economic growth Competitiveness	Macro-economic models of the real and monetary economies Strategic environmental assessment
National ↕ Regional/Urban	Economic development Regional competitiveness Regeneration	Forecasting models Environmental impact analysis Input-output models Geographical information systems Corridor and network assessment
Urban/Regional ↕ Local	Accessibility Liveable cities Distributional impacts Needs	Land-use transport models Project assessment Quality of life and other indicators Accessibility thresholds Minimum levels of service

Note
The issues and methods may operate at different scales but have been allocated the most relevant scale

focus of the means by which transport planning can be linked in with the changing demographic and industrial structure of society, the city as a place, and the distributional impacts of decisions. It is at these three levels that transport planning still has an important role in determining the quality of life in the new century.

Economic growth

One of the main criticisms made throughout this book has been that the transport focus is unnecessarily narrow as travel is dependent on land use and urban structure and the characteristics of the population, as well as the transport network. To produce a truly holistic view of transport demand it is necessary to integrate transport models with the broader perspective of macro-economics and the property market. Such an approach would not, however, be targeted upon the full spectrum of levels of policy interest in travel demand. It would be confined to the international, national and strategic regional levels, both of analysis and of policy formation (Table 9.4). It would not be so appropriate for local and urban scale analyses concerned with the details of network loading and mode choice, but should address the longer term questions of aggregate regional and sub-regional infrastructure requirements.

The argument for broader based approaches to strategic analysis is twofold:

◆ Travel demand is, in most cases, derived from other demands – from wishes or needs to participate in other social and economic activity. It is a paradox that, whilst this proposition is widely accepted, transport researchers have persisted in attempting to develop more or less comprehensive models which effectively detach the analysis of trip making from all but the simplest of socio-economic frameworks. Long-term travel demand changes should be analysed as derivatives and should therefore be modelled by drawing selectively from the fields of economic, demographic, property market, and land-use analysis, rather than autonomously as at present.

◆ Large-scale comprehensive models have not proved conspicuously successful in the past. They have been found expensive to develop, calibrate and maintain (due to their extensive data requirements and unwieldy structures), and have often been seen to exhibit implicit properties which tie their operability to particular phases of the long-term cyclical pattern of economic development. Classic examples have included Lowry's (1964) *A Model of Metropolis* and Forrester's (1969) *Urban Dynamics*.

The primarily derived status of travel demand becomes the lynch pin of a much looser modelling framework – one in which the outputs of major non-transport forecasting projects are combined in such a way that medium- to long-term shifts in national, regional and sub-regional travel demand can be modelled as derivatives speedily and without massive data gathering and calibration exercises.

Two such approaches can be identified which would link transport analysis with models of the 'real' economy and the 'money' economy.

Models of the 'real' economy

These models are based on macro-economic forecasts of the economy, together with its main regional subdivisions. Two macroeconomic standpoints are reflected:

♦   Three models (those of the City University Business School, Liverpool University Macroeconomic Research Unit, and the London Business School Centre for Economic Forecasting) are essentially monetarist in orientation, emphasizing the financial sectors of the economy and the links between money supply, interest, and exchange rates. They are typically based upon the assumption of national expectations and are all built using straightforward econometric procedures.

♦   The three other principal models (Cambridge Econometrics, the National Institute of Economic and Social Research, and the Centre for Economics and Business Research Ltd) have somewhat longer pedigrees and reflect a regression rather than monetarist economic disposition. They emphasize 'real' rather than monetary variables (demand, employment, and output). That of the NIESR is of a traditional econometric structure, but the Cambridge model is a latter-day derivative of Stone's multi-sector dynamic input-output model, much extended and embellished in recent years. The CEBR approach involves forecasting, the collection of surveys, and the development of statistical and economic analysis through econometric modelling, for example their study of the economic impacts of different levels of expenditure on the roads programme (CEBR, 1994).

Of these alternatives, it seems that the NIESR approach, based in a Keynesian structure, has most potential for travel demand analysis:

♦   It allows longer term forecasts over 5 to 10 years, whilst the monetarist models are short term with quarterly (sometimes monthly) forecasts over a 12 month period.

♦   A real economy model, which offers detailed sectoral disaggregation (for example, on employment and consumer expenditure), is close to a necessary condition for the production of regionally partitioned forecasts.

- A close match can be achieved between the disaggregate output measures above and those which are likely to influence demand, such as demographic factors. The regional balance between labour supply and demand (absolutely critical to the estimation of travel demand) can be established from publicly available sources.

- If the travel demand that is initiated by the regional economy is mediated through the pattern of land uses and the process of land-use change, then it becomes essential to translate economic forecasts into property market projections. Relative shifts in the commercial and residential property markets underpin the processes of land-use change, and in an increasingly unfettered way as the planning regime becomes ever more permissive.

Models of the 'money' economy

The case for relying heavily upon real economy models, stepped down to regional and property market levels, appears close to overwhelming. However, the partial position of the commercial and residential property markets in determining the land-use impacts upon travel demand does raise some further questions.

Regional and sub-regional property markets represent the outcomes, not just of user and occupier demand pressure, but also of independent supply side effects. Adjustments to rates of development are not just mechanical responses to shifts in rental values. The partial autonomy of the property markets is evidenced through their propensity, over the past 20 years, to move in cyclical patterns which have not matched those of the wider economy. At the very least there are significant time lags.

If it is the case that the supply side of the commercial property markets may get well out of step with underlying rates of demand, then millions of square feet of commercial space may get developed in a sub-region (for example, London) well in advance of what is required. The result is likely to be a collapse in rents which might, in turn, be sufficient to shift the underlying pattern of regional economic relativities. And the travel demand consequences over the period of 2 or 3 years could be very considerable.

Thus there does appear to be a case for considering the alternative economic modelling perspective which emphasizes monetary variables and through these the independent movement in the supply side of the various equations. The reasoning is as follows:

- The agencies involved include most of the major financial institutions (insurance companies, pension funds etc.), the property development companies, the house building agencies, and the main commercial owner occupiers who are forced to monitor the returns on all their assets including their property holdings.

- The capital investment and disinvestment policies of each of these classes of actors in the property markets are ultimately governed by the risk-return ratios prevailing in the international stock markets. Fund managers are constantly comparing investment in direct property with returns on equities (UK and non-UK), gilt-edged stocks and other long dated bonds, and liquidity (that is, return on the money-markets).

- Risk-return relationships are difficult to predict, but underlying their movements, the preliminary interest and currency exchange rates are critical. House builders and property companies, which together produce most of the country's new built stock, are predominantly debt financed, and so gearing ratios related to the costs of money are fundamental to their operations.

It seems, therefore, that a robust long-term travel demand modelling strategy would be based upon the use of at least two types of national economic model. The first would centralize the documentation of the real economy of demand, output and employment; it would be disaggregated to a regional and sub-regional scale; it would at this scale be linked with demographic models to balance labour demand and supply; and it would generate regionalized outputs appropriate to the calibration of property market models which are essentially demand driven. The second type of model would emphasize monetary variables and the supply side of the national economy; it would link the analysis of interest and exchange rate movements directly into the process of investment asset allocation forecasting; and it would provide independent development pressure inputs to property market models.

Comment

Macro-economic change and the impact of technology pose different problems of analysis as the scale is national and the effects occur over a longer period of time. Consequently, it is argued that the short-term local-scale analysis methods normally used in land-use and transport analysis may not be appropriate. It is here that macro-economic models together with models of the property market would be useful. Changes in demand for transport arise as a result of changes in the overall economy, in its industrial structure, and in technology. Keynesian models of the economy allow regional factors, demographic changes, and employment effects to be incorporated into the analysis for a time horizon of 5–10 years. Some of the approaches use an input-output model and can therefore explore the impact of industrial structure on location decisions.

On the other hand, monetarist models lay more emphasis on the money supply, interest rates and exchange rates, and thereby analyse shorter-term movements in some of the key supply side variables. Through investment asset allocation forecasting, flows of capital into the industrial and property markets can be estimated, and from this the levels of trip generation. Together, these two approaches could help in the estimation of total travel demand at the macro-level in the short and medium term, and transport analysis would be truly integrated with the economic and industrial structure at the international and national levels.

To supplement the various forms of economic analysis, it is also necessary to take a strategic view on environmental issues. Strategic environmental assessment (SEA) requires a commitment to sustainable development so that decision-making can a take a view that balances alternative priorities, including mobility, safety, environmental protection and economic development. These methods must also be capable of handling multi-modal issues, and address in a synthetic way measures that have both a direct and an indirect effect on the infrastructure. This is in addition to the macro-economic and property market effects outlined above. Effective linkages have to be built between transport and other sectors. Consultation with all the key stakeholders is an essential part of the SEA process. In all cases, the arguments for an against taking a particular decision should be presented in an accessible form. This means that methods should be transparent, with large amounts of data being summarized as a series of indicators on environmental impact and sustainability. Multi-layered maps, including GIS presentation, are useful in presenting the complexity, but it should be clear that there is a trade off between this complexity and a complete understanding of the underlying issues and trends. Their purpose is to assist in a high-level decision-making process. Data driven assessments can mislead, particularly when considering long-term outcomes, as they mask uncertainty (ECMT, 2000).

The promotion of economic development

Planning has switched from the notion of directing growth through investment to that of promoting development. The infrastructure necessary for development is no longer provided at public expense so that developers can enjoy the benefits through higher rents. An essential part of the development process is that of negotiation between the public and private sectors through exploring the means by which cooperation can take place. Previously, planners had been engaged mainly in designating land uses, and in securing adequate public infrastructure, amenities and housing. The concern was over comprehensiveness and long-term strategic objectives together with the ideal that development could take place with the interests of all parties safeguarded. This ideal has long been discredited as the interests of particular groups (for example, business and car drivers) are always promoted over those of others. Altshuler (1965) argued that planners only had limited knowledge and this made their understanding of the public interest problematical, whilst Harvey (1978) and Foglesong (1986) both concluded that the planning system was oriented towards capital accumulation and that diversion into equity considerations was limited to further legitimation of dominant economic and political interests.

The industrial restructuring of city centres and the decentralization of work and homes to greenfield sites have caused a radical reassessment of the role of planners as regulators. There is now a direct connection between the economic structure and planning legitimation (Fainstein, 1991). Ideology no longer obscures the planner's role and the difficulty now relates to the means by which benefits to private developers can be equated with public sector objectives. The arguments used by governments concerning the efficiency, accountability, and productivity of private firms have not been countered by an effective alternative argument that activities traditionally carried out in the public sector should continue to be exclusively in the public domain. Partnership is essential, and the balance has switched to the private sector.

In the United States, planners have either been political agents, or have worked in public development corporations or in private consultancies. Formal procedures are not followed but there is a process of continuous interaction (Fainstein, 1991). Subsidies can be offered to industry willing to remain or expand in the city centre, tax policy can be manipulated, regulations can be loosened, infrastructure can be built, and deals negotiated to allow developers to trade public benefits for mitigation of regulations. The British system is more rigid with set procedures and a wide ranging documentation of strategies at the county and local levels.

However, even here greater flexibility is being encouraged, particularly at the interface between the public and private sectors. The development orientation allows planning gain and in certain situations the planning system has been bypassed. Urban development corporations, then city partnerships and urban regeneration companies, have been set up to facilitate rapid development, including land assembly, tax relief (originally in enterprise zones, then partnership areas and brownfield sites), and in principle to speed up the necessary associated infrastructure construction. Perhaps British planners need to be given comparable powers to their counterparts in the United States to allow a much greater flexibility and powers of negotiation so that industrial and commercial development can be attracted to city locations. Rather than setting out a series of long-term objectives together with the means to achieve those objectives, planners are now concerned with action planning. This involves more limited objectives, and an overriding concern with seeing things happen and seeing opportunities as they occur. Beauregard (1990) calls this process city building, where narrow goals are set and all the means available are used to permit progress. This is the city competitive based on market rationality. Notions of long termism, a rational comprehensive approach to planning, and the minimization of externalities become irrelevant when objectives are short term and involve negotiations with the private sector.

Cities are now looser agglomerations with the link between homes and workplaces being broken. Industry is also footloose with location being based on a wider range of factors including the means to reduce costs through relocation. The functions of cities are no longer based on manufacturing production but on the supply of services (for example, retail, financial and educational). Globalization of national economies and the close interrelationships between financial, commercial and manufacturing markets have all led to fundamental economic restructuring based on the new generation of communications, database and information technologies.

One of the major unresolved research issues in transport is the question as to whether transport infrastructure investment promotes economic growth at the regional and local levels. The concern is not with the transport benefits, principally measured as travel time savings, but whether there are additional development benefits from these investments. If they do exist, can they be measured? New approaches are now being developed based on defining the set of necessary conditions for economic development to take place (Banister and Berechman, 2000 and 2001). In addition to the economic conditions, there are the investment conditions and the political and institutional conditions. Only when all three sets of necessary conditions are operating at the same time will measurable and additional economic development benefits be found, and that the most appropriate modelling framework would be based on micro-economic approaches, together with simulation studies (see Chapter 5).

Comment

It is within this dynamic continuum that planning must be placed. There are now many different means to achieve a particular objective. This can only be realized through shorter-term market analysis, project and programme evaluation, and a negotiation process. Planners become trapped in the logic of the bottom line, trading off regulations for private sector commitments to stay or build or contribute to community betterment (Fainstein, 1991). Within this process it is impossible to separate transport from the development process as it has played an integral part in allowing the deconcentration of cities. The evolving land uses have increasingly become car based with little opportunity for shorter journeys where walk, bicycle and bus would be more appropriate.

Newman and Kenworthy's (1989 and 1999) global analyses of energy consumption in different cities has concluded that car dependence can be reduced by physical planning policies, particularly re-urbanization and a re-orientation of transport priorities through measures such as:

- increasing urban density;
- strengthening the city centre;
- extending the proportion of the city that has inner area land uses;
- providing a good public transport option;
- restraining the provision of car infrastructure.

These physical planning policies need to be supported by economic policies on land use, housing and the provision of infrastructure. Hanson (1992) suggests an integration of physical planning, cost strategies, and coordination of land markets, housing markets, and transport systems if cities are to re-establish themselves as centres of economic development. At present, the incentives are encouraging more sprawl and low-density development which results in car dependence and a high use of resources. Cities have to be seen as attractive places in which to live and work, and transport planning has an instrumental role to play in achieving that objective.

Environment and health

Perhaps the greatest change in transport thinking has been the growth in interest in the environment and the possibility (or probability) that there are direct links between transport and health. There is no doubt that transport affects the quality of life, both positively through access, recreation and exercise and negatively through accidents, pollution, noise and vibration, stress and anxiety, severance, loss of land and blight (Whitelegg, 1993). The new debate is over the effects of transport on health, and how pollutants can directly (or indirectly) affect that health individually or in combination. Powerful arguments have been raised by economists (Maddison *et al.*, 1996) and the medical profession (BMA, 1997), which have added to the more established evidence about the negative impact of the car on community networks (Appleyard, 1981), the reduction in walking and cycling (Hillman *et al.*, 1990), and the declining (yet still high) levels of road accidents.

The economic case is that pollution is a transport externality and should be priced so that the polluter pays. The difficulties of measurement and monitoring have been overcome through the use of satellite imagery, as is being proposed in Singapore. This will enable point sources of pollution to be identified and measured as vehicles move through the road system so the individual polluters can be recognized. The real difficulty comes in allocating a cost to this pollution in such a way that its environmental and health effects are internalized. Maddison *et al.* (1996) concluded that 'on average, the external costs inflicted by the last road user amount to between 11.2 pence and 12.9 pence. The tax paid per kilometre is on average only 4 pence' (p. 141). This calculation has been based on the best available information and some strong assumptions on the general linkages between pollution and health, on dose response effects, on the direct and indirect impacts on morbidity, and on the links between pollution and premature death.

The medical evidence is also problematical, as it is difficult (or impossible) to establish quantitative links between the effects of transport policy on health. There is a wealth of epidemiological information and a larger amount of transport data, but the difficulty arises in establishing clear statistical relationships, and then to infer some causality. Health is influenced by socio-economic conditions, diet, housing condition, lifestyle, family history, and many other factors, including transport. The BMA (1997) in their policy recommendations concentrated on reducing reliance on and the need for health damaging forms of transport, increasing the use of health promoting forms of transport, increasing mobility and access and reducing inequity, and reducing the negative effects of modes of transport. Such a focus avoided the difficult questions of attributing cause and effect. It does not attack the cause of the problem, but looks more towards damage limitation through promotion of alternatives, and seeking to reduce the use of the car through pricing, restraint and technological improvements.

Comment

The agreement here is clear, namely that even though the science is not well known, the potential impact is substantial, and that the precautionary principle should be adopted. It is not just the medical evidence that is important here, but also the psychological impacts on individuals, the general quality of the urban environment, and the potential contribution of healthy transport to sustainable development. The effects are also likely to impact more seriously on low-income city dwellers in poor quality housing, and upon the elderly and the young, principally demonstrated through the huge increases in levels of childhood asthma. Analysis of deaths in France, Austria and Switzerland shows that 6% of all deaths (about 40,000 a year) 'stem from air pollution, around half due to tiny particles in vehicle exhausts, particularly diesel'. The *Lancet* report goes on to state that 25,000 new cases of chronic bronchitis in adults, 290,000 cases in children, and more than 500,000 asthma attacks can be attributed to transport (Künzli, 2000).

Planners have new responsibilities, as set out in the Environment Act 1995 (effective from 1 April 1997), to address the problems of identifying pollution 'hotspots' and to establish action plans. These action plans have to be developed for designated Air Quality Management Areas where air quality objectives are not likely to be met in 2005. Traditional traffic management measures are not likely to achieve these objectives on their own, as the problem may migrate from one location to another and as measures to reduce congestion have often resulted in longer journeys. Technological solutions, pricing and demand management may all help, but planning can also have an instrumental role to play through transport action plans. These must examine the means to reduce demand for motorized travel. This primarily means that local authorities must place restrictions on the use of cars, take positive measures to promote growth in the use of green modes and public transport, and examine area wide schemes to improve the liveability of urban areas.

Demand management measures should determine the most suitable forms of road use in the urban area (Table 9.5). Access control measures can be introduced to ensure essential users, residents and others have priority access to the area, whilst traffic calming measures reduce speeds and limit capacity. The key question (still unresolved) is whether existing traffic is diverted or whether it is degenerated (or disappears). Management measures which only bring benefits to particular locations need to be carefully assessed in the broader environment as the problems may again migrate to other locations. If this takes place, then an assessment of the air quality impacts should form part of the overall decision. Residential air quality may be more important to improve than air quality in a commercial area. Similarly, park-and-ride schemes offer an opportunity to reduce the use of the car in the urban area by encouraging switching to public transport. As with most options, park-and-ride needs to be combined with parking restrictions in the town centre and priority to public transport through exclusive rights of way and priority at junctions. High quality public transport can be combined with less use of the car in the urban area to achieve air quality benefits.

The promotion of public transport must be seen as an essential positive element in any transport action plan (Table 9.5). The image of bus services in towns and cities has improved substantially with investment in new vehicles, particularly midibuses and minibuses. But only limited priority has been assigned to them through bus only lanes and traffic lights. Air quality targets allow more radical options to be considered including separate rights of way for buses in town centres and along the radial routes into town centres. Bus only roads and routes will make journey times faster and help promote bus use.

Emissions levels relate to the vehicle, so an empty bus produces more pollution than a car, but a full bus (40 people) produces only 17% VOC and 56% CO emission levels of a full car (4 people), but still 4.5 times NO_x levels

Table 9.5 Transport action plans

Demand management	Access control measures; traffic calming measures; park-and-ride
Promotion of green modes and public transport	Bus only roads and routes; cycle routes, pedestrian priority; hybrid and electric vehicles.
Area based strategies	Clear zones and areas – only accessible to clean vehicles and green modes (cycle and walk); pedestrian friendly zones; car free housing; home zones

and particulates (DOE/DOT, 1996, Table 3). Occupancy levels must be a key component of all air quality strategies, for cars and for public transport. But the arguments for public transport are not so clear cut in pollution terms as often suggested. As bus size has decreased and as current occupancy levels in buses are about 12 people, the advantages in emission terms are limited or non-existent (the current average car occupancy levels are 1.67 persons). Table 9.6 shows that at current occupancy levels, the bus produces 6.4 times as much NO_x as the car per passenger-kilometre, but only 78% and 36% of the CO and VOC emissions, respectively. By giving public transport clear priority, the amount of stop-start use can be minimized and operating conditions improved. If car users switch to public transport there will be reductions in emissions levels, but only if those cars are not put to alternative uses. It is only when new types of fuel are considered that the bus may at least gain clear emissions advantages over the car.

In the short term the most important single measure to meeting air quality targets is to encourage cycling and walking, both of which are non-polluting forms of transport. The national cycling strategy (July 1996) aims to double the amount of cycling over a 6 year period (to 2002). This is a fairly modest

Table 9.6 Emissions levels for cars and buses in urban areas

Vehicle type	NO_x	PM_{10}	CO	VOC
	Exhaust emissions (g/km)			
Car (1 person)	0.31	–	3.37	0.23
Bus (empty)	14.46	0.74	18.90	0.57
	Exhaust emissions (g/passenger-km)			
Car (4 persons)	0.08	–	0.84	0.06
Bus (40 passengers)	0.36	0.02	0.47	0.01
	Exhaust emissions (g/passenger-km)			
Car (1.67 persons)	0.19	–	2.02	0.14
Bus (12 passengers)	1.21	0.06	1.58	0.05

Notes
Car is a post 1/1/93 petrol vehicle with a catalytic converter.
Bus is a post 1/10/94 diesel vehicle. The occupancy levels are typical of the current levels of use.
Sources: Banister (1997b), and DOE/DOT (1996).

target as the amount of cycling has reduced by 25% over the 10 years 1985–1995. To achieve more challenging increases for both walking and cycling, priority must be given to pedestrians and cyclists through exclusive rights of way, through direct routing, and through essential support measures (for example, safe storage of cycles). Cyclists could also use bus only roads and routes. Other European countries have been far more active in promoting quality facilities for cyclists and pedestrians, and this is reflected in the higher proportion of trips by these modes (Tolley, 1997). However, as with bus travel, to achieve real air quality benefits and the double dividend, encouragement to switch modes must be given to car drivers. There are benefits from the increased use of public transport and cycles, but the double dividend is achieved only if the car is not used at all.

Perhaps the most interesting opportunities lie in the area based strategies (Table 9.5). One way to proceed would be to establish clear zones or areas within urban areas where only non-polluting modes of transport would be allowed. Electric vehicles would be used within the area, with all polluting forms of transport being parked at the periphery. These areas would provide a clear demonstration of the benefits of zero levels of transport pollution and the quality aspects of city living would be considerably enhanced as these locations would be clean and quiet areas. Alternatively, car free housing could be introduced or pedestrian friendly areas where cars are only tolerated under special conditions. The achievement of challenging air quality targets in our towns, through a mixture of strategies such as those suggested here, may provide the catalyst for a revitalization of cities and city living. This discussion is developed further in the next section.

Cities for people

Cities are becoming less accessible to many people. This deterioration is likely to continue with further dependence on the car and higher levels of mobility, both of which will lead to severe congestion as the capacity of the transport system fails to respond. Problems are created for industry, the environment, and for people trying to lead fulfiling lives. Cities then become less attractive places in which to live as decentralization takes people and jobs to peripheral car-accessible locations.

To reverse these trends cities must become attractive with affordable housing and high-quality facilities and amenities. There is a real urgency for action in the UK, and in the South-East in particular, where there is a danger that that region has become dysfunctional. This means that the economic base is fundamentally weakened as there is an emerging disparity between the rich who have a full range of job opportunities, housing, a high-quality lifestyle, with high levels of mobility, and those who do not. This second group, including many public sector workers in schools, hospitals, the police and other support services, is essential to the operation of the economy, yet they do not have access to the same range or quality of opportunities. They have been effectively priced out of the market, yet the market depends on them for its existence – this is the dysfunctionality. There is an urgent need for wage levels to be reassessed and for affordable housing for all workers, and cities must be seen as safe and pleasant places to live. Transport plays a pivotal role in achieving such a city, and although the car may still be essential it must be seen as promoting cities rather than as one of the main reasons why cities are perceived as hostile environments.

The car. The car is an inefficient user of road space and there is a range of options available to increase the efficiency of use of urban road space. These include road pricing, the use of smaller vehicles, and ensuring greater time in use because when a car is parked it is still using valuable space. Road pricing is now a real option, but there are still questions over its implementation which require much more general public acceptance than is now apparent. There are also major questions which have not yet been resolved, including the distortions caused by company cars, boundary and land-use pressures, land values and rents, the impact on the economic viability of city centres, the geographical area of application, equity and distributional issues, the use of revenues raised from road pricing, and even the impacts on the car industry.

Smaller city cars and more intensive use of taxis are a real possibility, and several manufacturers are exploring lean burn engines (for example, Ford), electric city cars (for example, General Motors and Fiat), hybrid vehicles (for example, Toyota and General Motors have produced production versions of hybrid petrol/electric vehicles), and the use of alternative fuel sources including fuel cells (for example, Daimler-Benz, Ford and Ballard Power Systems as a joint venture have several types of fuel cell vehicles). Provided that these vehicles are given appropriate tax advantages (for example, no vehicle excise duty) and priority (for example, privileged access to parking spaces), then there would be sufficient incentive for their purchase.

Alternatively, the Californian requirements for 2% zero emission vehicles by 1998 for companies producing more than 35,000 cars (increasing to 10% by 2003) gives clear incentives to industry to develop cars more appropriate to city use. Although in this case, the requirements were too difficult to meet and so the time horizon has been delayed for three years to 2006. Companies not developing their own zero emission car can sell another company's car at a subsidized price and cover those losses through raising the price of conventional petrol cars. Similarly, if a company sells more than the magic 2%, credit can be claimed and those credits can then be sold on the open market to other companies to fulfil their quotas.

Parking. Similarly, a market priced parking strategy which involved appropriate enforcement measures and controls over all forms of city parking would
ensure minimum parking time and full costs imposed on users. These parking
charges would be equivalent to the office rent and unified business rate level
for the particular area. This would mean a doubling of existing hourly parking
rates in Central London (Banister, 1989). Effective parking policy must include enforcement and payment of fixed penalty notices together with the
identification of frequent offenders, whilst removal of vehicles has acted as a
major deterrent to illegal parking (NEDC, 1991).

Recent measures in the Transport Act 2000 have facilitated work place
parking levies, but not on other forms of parking (for example, shopping).
Local authorities will be able to use the revenues raised to pay for public
transport improvements, as well as some road investments. This is the first
time that hypothecation of transport revenues has been achieved, but at the
cost of making its implementation a local responsibility rather than a government one.

The bus. The bus is the most efficient user of city road space, yet it has to share
road space and this in turn reduces its efficiency. The urban road network
could be designated for particular users (see Chapter 8). A central shopping
area would have a higher pedestrian allocation, while a peripheral office
location would have only a limited pedestrian and public transport allocation,
and a higher general allocation. The implementation of such a scheme could
take place immediately, and the capacity of the bus system would be
dramatically increased as scheduled operating speeds would be improved.
There would be fewer cars in the city centre and the environmental benefits
would also be significant. The real value of such a scheme would only
materialize if a network of bus only routes was introduced and buses were
completely separated from cars on 'Green Routes'. The 'Red Routes' in
London are a partial step to such a position, as parking prohibitions are strictly
enforced, allowing the free flow of traffic. But buses still have to share road
space with other traffic. The 'Green Route' option is now being considered in
London in the context of bus deregulation as a set of exclusive rights of way
would probably encourage greater on-street competition between operators.

It is not just the physical infrastructure that needs addressing, but the
quality of the bus service more generally. This means better information to
the user, integration in the widest sense, and a system that is easy and attractive
to use. There is a tendency to concentrate on the larger investments, such as
those in new light rail transit systems, but equally effective are improvements
in the basic unit of public transport, namely the bus in its many forms.

Coordination between modes and high occupancy vehicles. As a part of this priority strategy, coordination and integration between modes must be promoted, so that the necessity to bring a car into the city centre is reduced. Parkway stations and park-and-ride facilities can be located at peripheral sites or at suitable interchange points at accessible sites within the city. High occupancy vehicle lanes can also be introduced on main motorway routes into city centres to encourage car sharing. Firms can positively encourage car sharing. This possibility is central to most of the current generation of company transport plans, where flexible forms of car sharing have been set up together with incentive schemes to encourage car drivers to share or to switch modes.

Transport solutions are multi-modal and involve increasing the occupancy levels for each form of transport. This has clear environmental advantages, as well as making the best use of available network capacity. Consequently, the financing of new investment and the subsidy of transport should be seen in the same way. This means a move away from single mode financing towards decisions, which relate to improvements in transport quality and reductions in congestion. The multi-modal studies (DETR, 2000b) are a move in this direction, but there is an underlying suspicion that the outcomes will still result in more roads being built, rather than a truly integrated transport system being developed.

Capital investment in urban public transport. In addition to the management of the existing road space and the allocation of space to priority users, capital investment is still desperately needed to update the existing public transport infrastructure. Several Light Rail Systems (LRT) are now being constructed in British cities with the initial phase of the Manchester Metrolink (1992) providing the first example of the new generation of schemes. Schemes are now in operation in Manchester, Sheffield, Newcastle, the West Midlands and Croydon. Larger scale investments are planned for London with the Cross Rail scheme, extension to the Docklands Light Railway (to City Airport), completion of Thameslink 2000, and the Wimbledon-Hackney underground. In all of these proposals, the financing has proved difficult as the government has (in the past) imposed strict limits on public expenditure and as private finance has not been made available in the required amounts. In most LRT schemes, the local authority has financed the infrastructure costs together with Section 56 Grants (Transport Act 1968) from central government, and the private sector will be responsible for running the services.

With the commitment made in the Transport 2010 document (DETR, 2000b), substantially increased amounts of public expenditure have been made available, but this is contingent upon the private sector also investing. This provides an important new opportunity for capital investment in public transport, but the requirement for matching funding in many cases may result at best in limited investment, and at worst no investment. As discussed earlier, the public sector must provide most of the capital investment requirements, and this fact needs to be recognized. The role of the private sector should be mainly limited to running services under contract, and in providing some supplementary investment. If total reliance is placed on the private sector to provide substantial financial inputs, there will be further delays in essential new and replacement investments. This is the public service challenge accepted by the new Labour government (2000*b*), and it will be judged on its ability to deliver.

Allocation of city centre road space. Allocation of city centre road space to particular modes may seem a radical proposal, but in other respects this strategy is the only one available in the short term, and it has been adopted in part through pedestrianization and traffic calming. Traditionally, the case for both these options has been argued on safety and (more recently) environmental criteria. Congestion has not featured. Unless pedestrianization and traffic calming are carried out on an area wide basis, they may just divert traffic away from one area to another, redistributing traffic but not reducing congestion. The environmental benefits are not convincing as research has not been able to establish a link between traffic speeds and noise and pollution in the urban environment. Most evidence relates to residents' perceptions of improvements in environmental quality when traffic levels and speeds are reduced.

Traffic calming schemes are now accepted and many are being implemented with a high degree of public acceptance, even though many of these schemes were not implemented imaginatively or in terms of an area wide strategy (Environment and Transport Planning, 1991). For transport planners, there seems to be a series of related problems in implementing these strategies at the city level, and the problems are well illustrated with respect to the issue of traffic calming and more recently home zones. First, traffic calming only rarely seems to be introduced as part of a coherent transport strategy. Each individual application is dealt with separately on its own merits. There seems to be a clear demand from residents to 'calm' their local areas, but there is no established methodology to achieve a priority ranking for all possible applications. Local authority budgets are limited and there is a concern among both politicians and professionals that any implementation strategy should be fair. Where attempts have been made to assign priorities to different schemes, the political implications have resulted in their rejection. For example, in the Royal Borough of Kensington and Chelsea (London) priorities were allocated on the basis of traffic volumes, traffic speeds, accidents, aesthetic values, costs, and the size of the benefiting population. However, the results were not acceptable to the local politicians whose own local knowledge was at variance with the results of the method developed by the consultants (Frank Graham Consulting Engineers, 1991).

The second related problem is the impact of the traffic calming scheme on the urban area as a whole. Positive responses from those living in the traffic calmed area are more than outweighed by anger from those living in adjacent areas where traffic levels (and accidents) have increased. Traffic calming must be seen as an area wide traffic management strategy as traffic and accidents may just migrate to neighbouring locations. The debate on this problem is not clear-cut as some (for example, Tolley, 1990) argue that traffic actually disappears and crucially does not appear elsewhere. The circumstances under which this interesting proposition might be supported need further investigation – 'building roads generates traffic, removing them degenerates traffic' (p. 115). Tolley's vision (1990) would be to use transport planning to reduce trip lengths, to encourage a switch to soft modes (walking and cycling) and public transport, and to 'domesticate' the private car.

Other recent initiatives look to the local community to propose 'living neighbourhoods' where partnerships are built between the community, local authorities and service providers to find ways to reduce car use. This sort of approach is an extension of the 'travel blending' concept where again the local community is involved in finding specific local measures to address the problem of the car so that people can better understand their own options. Apart from seeking solutions through sharing, trip tours and a more imaginative use of transport, these approaches begin to empower people through bringing together all parties in discussion and eventually a partnership. The professionals can also learn from this dialogue. At a higher level, it reflects the move towards 'green transport plans' now being introduced by many private and public companies or in particular locations, such as technology parks and airports. Extensive debates lead to plans and targets being set to reduce the use of the car for the journey to work and other facilities. The problems of the car are well known and all people are part of it. They should all be part of the solution.

Integrating land-use and transport planning.   Perhaps the most important contribution to reduction in congestion is to acknowledge the close links between land use and transport, and to make a greater use of the planning system to control transport growth. The principal objective must be to maximize accessibility and minimize trip lengths. These twin policy objectives would guarantee the greatest levels of demand for public transport, cycling and walk modes. By ensuring the appropriate mixture of land uses, the availability of local facilities and employment, and good quality public transport, the greatest efficiency in levels of transport would be obtained and the levels of energy consumption per trip and per person would be minimized, even where trip generation rates are high. The environmental costs of transport would also be significantly reduced if the dependence on car travel was reduced and balanced communities were encouraged (Banister, 1992*a*).

In the Green Paper on the Urban Environment, the EC argues very strongly for a compact city as the solution to the problem of urban congestion, lower energy consumption and lower pollution levels, and for the improvement of the quality of life (CEC, 1990). It goes beyond the sole concern with environmental sustainability to cover the impacts on the natural environment and the quality of urban life, and it views the city as a resource which should be protected. The quality of life in European cities has deteriorated as a result of uncontrolled pressures on the environment. In addition, the spatial arrangement of urban areas has led to urban sprawl and the spatial separation of functions. These factors have undermined the compact city which in turn has reduced creativity and the value of urban living. The return to the functionally-mixed compact city is a solution, the only solution to these problems.

This does not mean a return to the high densities of the city of the early twentieth century, but cities with a medium level of density (between 30–40 dwellings per hectare). The Urban Task Force's report (1999) proposed a vision for the urban renaissance that promoted cities that 'are well designed, more compact and better connected . . . allowing people to live, work and enjoy themselves at close quarters within a sustainable urban environment which is well integrated with public transport and adaptable to change' (p. 2). This message was reinforced by the Urban White Paper (DETR, 2000*c*), with its support for this vision and proposals for better planning and design, bringing previously developed land and empty property back into beneficial economic or social use, and looking after the existing urban environment better.

At present, when decisions are made on whether to approve new developments, traffic generation forms an important element but only on a project basis. Assessment of the transport implications on the system as a whole for the complete range of individual projects does not take place. A comprehensive perspective would allow transport impacts to be assessed on the plan and programme levels, not just on the individual project level. Transport is a key factor in influencing land-use and development patterns and it is also linked closely with economic growth. This applies to freight and passenger transport. This means that a transport assessment should be seen as an integral part of any major development, and developers should take on the responsibility to ensure access is possible by a range of transport modes. This thinking is now being reflected in the revised PPG13 (DETR, 2001) and in the requirements for the new local transport plans to take a view that combines transport with sustainability objectives. There is also an increasing involvement of many employers (in both the public and private sectors) in the development of company transport plans. New thinking does seem to be evolving, if slowly, and many of the right questions are now being asked. What is more promising is that some of the answers are also now being implemented.

Comment

Clear links have now been drawn between transport congestion, the environment and the quality of life. All the evidence leads to one conclusion, namely that cities must develop as attractive, safe places to live, and they must provide affordable housing, employment opportunities for a range of skills, and a diversity of social and leisure opportunities. Policies can all operate in the same direction to improve the environment, increase efficiency and promote accessibility. Any solution needs to combine a range of planning, economic and technological policies into packages which are appropriate for the case under study. There is no universal solution, but good implementation in one location does help 'sell' it in another, and learning from experience (both good and bad) is informative (Banister et al., 2000).

Traffic calming and pedestrianization are seen by some as the only feasible solution to the problems created by the car (Tolley, 1990). Limitations on the use of the car through attempts to slow it down result in enhanced safety and improved quality of life in urban areas. It is not possible or desirable to continue to build roads to accommodate the expected growth in the demand for road traffic. Increases in capacity encourage more travel and longer trips, and reductions in capacity may encourage less travel and shorter trips.

Promotion of coordinated public transport on exclusive rights of way (rail and bus routes) would provide fast, reliable and safe services and help change the traditional image of buses. One of the main benefits from deregulation has been the growth in minibus operations. In 1985–86 minibuses and midibuses accounted for 14% of the British bus and coach stock. By 1998–99 this figure had increased to over 32% (25,400 vehicles). These smaller vehicles fill the gap between the taxi and the large bus, they allow hail-and-ride operations, they increase the feeling of a personalized service, and they provide greater security to the traveller (Banister and Mackett, 1990). Flexible routeing, shared taxis, and the greater use of technology through passenger information systems can all increase the attractiveness of public transport.

Congestion pricing and parking controls are 'auto equalizers' designed to remove built in biases and subsidies (Cervero and Hall, 1990), but in themselves will not resolve the problem of urban congestion. Many of the problems of congestion relate to non-recurring incidents such as accidents, lane closures and unpredictable events. In the United States, some 50% of freeway congestion is caused by these types of incidents (Cervero and Hall,

1990). It is here that technology has a key role to play through driver information systems such as route guidance and variable message signs. Both of these systems give drivers real-time information on traffic conditions. Continuous monitoring of roads allows emergency vehicles to reach incidents in the shortest possible time, it can help minimize the length of any delay, and it can help slow traffic and redirect if necessary. In the longer term, fully automated highways and small electric/hybrid city vehicles may replace the current generation of cars. These vehicles may be hired or leased rather than owned, and would complement other more traditional petrol/diesel cars which would be used out of the city or on non-automated highways.

Land-use policy can also play an important role in the integration of public transport and land use, in ensuring that jobs and housing opportunities are in balance, and in designing people-friendly environments. If these objectives can be achieved, journey lengths would be reduced and the potential for using environmentally friendly modes (bus, cycle and walk) increased. Higher density mixed use developments in cities will enhance accessibility. It is in the suburbs that the increasing level of congestion is greatest and the solutions seem less clear (Cervero, 1989). The suburb to suburb movements on circumferential routes are ideal for the car, but the road capacity is limited. One option is for new road investment, but there is likely to be considerable opposition to new routes within existing urban areas. Public transport alternatives are less attractive, as are the opportunities for road pricing. The main policy lever available is land-use controls which provide for mixed developments, which will reduce trip lengths, but even this option will not eliminate congestion. If the example of the United States is taken, congestion within the city is followed by similar levels of congestion in the suburbs and beyond.

Meeting social needs

One of the major roles for planning is to ensure a fair distribution of resources according to clearly established political priorities and that this distribution allows all people to gain access to the facilities which they need. The problem with a car-based society is that not all people have access to a car, and even those who do have a car in the household may not be drivers or the car may not be available. By examining the travel characteristics of selected social groups it is possible to assess some of the variations.

In a comprehensive review of UK national transport policy for the period 1960–1990 it was found (Hay, 1992) that there had been relatively little reference to equity, fairness and justice (EFJ) by either of the two main political parties, and EFJ issues were more often brought forward by minority pressure groups within the party or by backbenchers in Parliament. Of a wide range of definitions relating to equity, fairness and justice, only two were widely used. Formal equality was raised with reference to who pays the costs of providing transport, and basic need was raised with reference to the provision of public transport. The former concept requires that like individuals should be treated in a like manner ('horizontal equity' in Table 9.7), and the latter is the principle that people should be able to secure the minimum required to sustain life and health ('equalization of opportunity' in Table 9.7). The use of EFJ concepts has widened since 1978 (Trinder *et al.*, 1991). In a similar study based on interviews (1989) carried out in the shire counties and metropolitan boroughs, similar results were found. Meeting basic needs and recognizing formal equality feature dominantly with both councillors and officers (Hay and Trinder, 1991).

It is clear that there is a wider recognition of EFJ issues among both politicians and planners, but there is less agreement on what should be done because neither party is certain how to tackle the problem in the transport sector. It is also clear that all substantive transport planning proposals will raise EFJ issues.

One of the key responsibilities of transport planners is to understand and resolve conflicts between interest groups, and the understanding of EFJ must be an integral part of those professional skills (Hay, 1992). It seems that a narrow view of EFJ has been taken at both the national and local levels. Here an attempt has been made to combine various elements of EFJ through the two principal definitions of formal equality and basic needs (Table 9.7). These

Table 9.7 Equity and transport

	Equalization of outcomes	**Equalization of opportunity**
Horizontal equity – Formal equity	Service distribution according to demand Service provided according to commercial criteria Market based with no subsidy	Equality in service distribution Service provided to all communities at a similar level Standards or minimum levels of service
Vertical equity – Substantive equity	Service distribution according to need Travel concessions for elderly, young, disabled and other need groups Subsidy provided to the user	Positive discrimination for particular disadvantaged groups Special services Subsidy provided to the service

Note
It should be noted that this table extends the interesting work of Rosenbloom and Altshuler (1977) who identified three main concepts of equity as they apply to urban transport:
- Fee for service – to each according to his or her financial contribution. Service distribution according to demand.
- Equality in service distribution – to each an equal share of public expenditure or an equal level of public service, regardless of need or financial contribution.
- Service distribution according to need – to each a share of public expenditure or service based on need, as government has chosen to define it and taken steps to ameliorate it.

concepts are supplemented by those of substantive equality and procedural fairness, together with various notions of needs (for example, need as demand and wider needs).

Equity can be conceptualized as an equalization concept which is concerned with the equalization of opportunity and outcome. Two further elements can be added to this basic dichotomy. Horizontal equity requires government to treat like persons alike in decisions concerning funding, the distribution of benefits and compensation. However, as Ellickson (1977) comments, likeness is a matter of degree and he suggests that horizontal equity can be breached where Michelman's test applies (1967): this requires a person to bear a loss and that this is not unfair 'if he should be able to perceive that a general policy of refusing compensation to people in his situation is likely to promote the welfare of people like him in the long run'. Vertical equity relates to the fairness in the distribution of wealth among different income groups, and as with equalization of outcome requires some assumptions on income distribution.

In effect, it is argued that governments plan public projects either to stimulate some measure of income redistribution or to improve the relative position of the disadvantaged through the equalization of opportunity. Table 9.7 summarizes the matrix of equity considerations. It should be noted that decisions in transport bear on each of our four components of equity, but in no coherent manner. With present policy, it seems that commercial criteria will prevail over other considerations, and this is reflected in its increasingly dominant position with the reductions in subsidy and competition in the provision of transport services.

The apparent simplicity and tidiness of the table conceals a much more complex reality. However, by presenting a flexible yet strong framework for the identification of the disadvantaged and an evaluation of alternative forms of transport in terms of the outcomes and opportunities available, a clear picture of the importance of needs based transport planning can be presented. One major role for politicians and transport planners at both the national and local levels is to set the agenda for transport based on concepts of equity. A theoretical framework such as that developed here provides the context, and service provision can be based on assessments of the needs for the various disadvantaged groups (Banister et al., 1984).

Else and Trinder (1991) interviewed local government councillors and officers to establish the four major transport issues in their particular authority. Nearly 75% of the issues mentioned came within the infrastructure category (roads, railways, maintenance, traffic flow, parking, environment and road safety). A further 20% concerned public transport (bus, rail and the disabled), with the remaining responses covering finance, planning and social policy. The most important single issue was investment in the road network within the jurisdiction of the various authorities, and in many cases members and officers mentioned particular projects. The second issue was traffic flow (that is congestion), which, although related to the first, was seen as a broader problem for which new road infrastructure was not the only solution. The other two issues which scored more than 10% were bus services and road maintenance.

As might be expected, less importance was allocated to road construction and maintenance in Labour controlled authorities, with more importance being attached to public transport and the broader issues. But in all cases it seems that the predominant view was transport as a separate sector to which policy priorities were allocated in isolation, not as part of an overall strategy for the city or county.

Even in the latest policy thinking on transport (DETR, 1998a and 2000b), little is said on social inequality apart from the scale of the problem. It is acknowledged that 60% of the poorest 20% of households do not have a car, neither do 55% of those over 70 years of age. Women and those under 20 are seen to be most dependent on public transport. Lack of accessible and affordable public transport is seen to be a contributory factor to social exclusion, but no policies are specifically aimed at these important groups. It is presumed that they will benefit from the range of policy proposals being put forward, but nothing is said on how their transport difficulties can be overcome. The vast majority of the £180 billion package (DETR, 2000b) will favour the top income groups, and mobile car drivers. On spatial equity, the latest government thinking is clearer with some initiatives being taken to confront poor accessibility in rural areas (DETR, 1998a, 2000b, and 2000d). Behind the broad aim of equitable access to services in rural areas (including transport), is the need to develop sustainable rural communities with increased opportunity for all. The policy measures mainly involve increased subsidy for more flexible forms of bus services through the rural transport partnership being established in all rural counties. Particular importance is being attached to the potential for voluntary and community sector projects, and innovative services such as flexible taxi services. In many cases, the car or taxi is probably the most cost effective way of tackling transport need in rural areas, and the potential for rural car clubs and car sharing need to be revisited. Flexibility and innovation with minimum constraints would seem to be a way forward, and the new technology provides the necessary communications systems to provide a reasonable quality of service to those living in rural areas.

Comment

Social need must be placed much higher up the political and planning agenda, as many decisions have transport implications and consequences in terms of social exclusion. For example, one of the major transport trends in the recent past has been the increase in the use of the car and the growth in average trip lengths. A major part of this increase can be explained by the closure of local facilities and the trend towards increasing specialization within facilities, so that the nearest opportunity is not the one which will actually be used. These changes are taking place in all sectors, supported by strong arguments for economies of scale and scope with thresholds being set for minimum viability (for example, in schools). These rationalizations have affected schools, shops, banks, magistrates courts, petrol stations, hospitals and many other forms of employment and services. In all cases there may be significant savings to the supplier of the service, but no account is taken of the increase in user costs. The additional costs of travelling to work or to a facility are borne by the user, and in many cases this results in more car based travel. But there are also consequences for those without access to a car for whom journeys would become longer and more difficult to make.

It is not difficult to calculate the increase in these costs. If, for example, a local hospital is closed, the additional costs of travelling to the next nearest equivalent facility may be calculated. It is not surprising that longer journeys and increased car dependence are two major outcomes of such rationalization of services. But these additional costs do not form part of the decision as they are not costs which accrue to the supplier of the service. Recently, there has been much debate over the possibility of road pricing and significantly raising the costs of travel to the individual. It is argued that the costs of travel are too low and must be raised, particularly under congested urban and motorway conditions. There has been no debate over the potential increase in distances and costs incurred by the individual who may now be forced to use the car to go to hospital, to take children to school, or to reach to the nearest shops, nor of the consequences for low-income households.

One important role for transport planners must be to take a more holistic view of what is happening at the national and local levels so that the full implications of decisions taken in any given sector can be interpreted. There must be a balance between the narrow market driven criteria for rationalization of services and the broader society based evaluation. In some cases the broader arguments may be sufficient to alter the decisions to close a local shop, school or hospital. Substantial externalities suggest that the social costs may exceed the private benefits resulting from closure.

Unless some attention is paid to these broader planning issues, more travel will be made by car, trip lengths will continue to increase, and more resources will be used. Notions of accessibility are replaced by notions of mobility. The possibility of travelling to local facilities by non-energy consumptive modes is reduced and the ideal of sustainable settlements becomes ever more remote. A balance has to be found. The only way to achieve this is with an overall vision for urban planning and transport. Group Transport 2000 Plus (1990) concluded that 'the necessary restrictions on car use and parking must look beyond prohibitions or dissuasive tolls and enable full integration of private cars in the transport system . . . People are "trapped" into using cars' (p. 17). The real costs of increasing car dependence should be assessed as should the policies of reducing car dependence. At present, neither is.

Notes

1 The new Golden Rule means that Government will only borrow to invest and not to fund current expenditure. The Treasury has a Sustainable Investment Rule which means that net public sector debt as a percentage of GDP will be held at a 'stable and prudent' level, currently below 40% (House of Commons Treasury Committee, 2000).
2 The loss for SNCF in 2000 was over FFr1 billion, and it is its fourth loss in the last 5 years.

Chapter 10

Conclusions

What then is the future for transport planning? Over the last 40 years it has proved to be robust and resilient to change, and transport planners have remained loyal to the systems approach with as much as possible of the analysis being formalized in quantitative models. However, two major factors have now led to a reassessment of the traditional approach. These factors are not based on the limitations and critique of established methodologies, but on the radically changed nature of the problems facing decision-makers.

The first relates to the changes in the political environment with the ascendancy of the market view over the social and welfare perspective. The Anglo Saxon philosophy, which has been dominant in the United Kingdom for over 20 years, reduces the role for strategic planning and longer-term objectives, and adopts a commercial (or quasi commercial) approach to maximize the internal efficiency of the transport system. The market mechanisms are given priority, but it is still realized that there are potential market distortions (for example, monopoly power and externalities) which must be corrected by government intervention. Demand must be met at the lowest cost. This philosophy has become dominant in the United Kingdom, Denmark, Spain, Ireland, Portugal, Greece, Norway and Sweden. However the 'true market' philosophy has never really infiltrated deeply into the transport thinking and more recently it has become a 'social market' philosophy with a stronger role again for the plan led approach. Even the private sector has welcomed a clear framework within which to operate as uncertainty leads to greater caution in investment and a desire to reduce risk.

The alternative approach is the continental European philosophy where transport is seen within a wider social perspective and there is a *Droit de Transport*. Here, a longer-term view is taken of the role of transport in economic development and political integration, with clear investment priorities and a greater degree of master planning. Transport must form an integral part of the social infrastructure as it affects the distribution of population, employment, shopping, and social life, and it has a direct influence on health and the environment. This much higher level of government intervention is still the dominant philosophy in France, Germany, the Netherlands, Belgium, Luxembourg and Italy.

Currently, there is a move towards further convergence between these two views as more pressure is exerted on public budgets and as the politics of Europe continue to move towards greater integration and commonality in their strategic thinking. For example, in the transport sector more countries have introduced regulatory reforms within all forms of public transport through open access to the market, through privatization, and through tendering for services. There has been a tendency for governments to withdraw from the provision, support and regulation of transport services.

The second major change is the increased awareness that transport planning must be seen as an integral part of a much wider process of decision-making. Too often in the past transport solutions have been seen as the only way to resolve transport problems. This was first evident in the justification for more road building in the belief that a shortage of capacity was the main reason for congestion. Increasingly, it has become apparent that more capacity, particularly in roads, generates more traffic and allows for the dispersal of activities, which in turn makes the provision of public transport services problematical. Transport planning must be seen as part of the land-use planning and development process which requires an integrated approach to analysis and a clear vision of the type of city and society in which we wish to live.

All decisions influence this pattern of development, and if we are serious about achieving targets for reductions in emissions, improved safety, and a better quality of life in cities, then these issues must be fully reflected in the decision-making process. In many cases, only limited value seems to be placed on environment, accidents and quality factors. Yet increasingly, these are the issues about which people claim to be most concerned .

The transport planning process is still driven by the desire to reduce travel time and costs. The premise underlying this assumption is that people value quicker and cheaper travel. The irony here is that it is now being argued that travel is too cheap and that prices must be raised to account for other factors, particularly externalities. The importance of speed is also being questioned as drivers are being urged to reduce speed for safety and environmental reasons, and lower speed limits are now being introduced in residential areas and even on motorways.

So the two main foundations of transport analysis (time and cost) are being questioned, yet it is not possible radically to change travel patterns, at least in the short term. Simply raising the price of one fairly inelastic commodity (such as transport), for whatever reasons, merely means that additional revenues are raised as taxation for the Treasury. It is only when price increases are matched with quality improvements in transport services (particularly public transport), together with policies which make it possible to reduce the number and length of journeys by car, will any real benefits be achieved. It is not surprising that the public are against increasing the costs of travel if the quality of all services is poor and declining, and if there is no real way to alter their lifestyles as planning and development policies have encouraged more travel and greater reliance on the car.

Even within the transport sector itself, alternative modes should not be seen in opposition to each other and set up as mutually exclusive choices. Too little attention has been given to obtaining the 'best mix' of modes through multi-modal journeys involving interchange, or through exploring the opportunities for 'trip chaining' so that many activities can be carried out on one 'tour'. The new information technology should allow real time information to be made available to all users so that trip planning can take place in advance of a journey and so that flexibility can be introduced into a trip. Such technology should begin to make travel, in particular by public transport, a pleasure rather than a challenge. Similarly, interchanges should become places and spaces where people want to spend time shopping, relaxing or meeting other people. Railway stations and other interchanges provide a wonderful opportunity to create vibrant environments of a high quality, which in turn will attract people to linger and to make greater use of public transport.

Ethics in transport planning

The debate over the role that transport planning should have in the future is likely to continue, as is the artificial separation of planning from transport, the links between economic growth and transport, and the question of whether the costs of transport are too cheap. Throughout this book, it has been stressed that transport planning has a key role to play in forming an alliance with market forces at all levels. The perspective taken has been that transport planning is an agent for promoting economic development and for making cities attractive places in which people want to live, particularly for those who are seen as being disadvantaged (Chapter 9). Underlying this perspective is a need for professionalism in transport planning and the ethics of honesty. Transport planners should not try to claim that their methods and approaches to analysis have all the answers, but they should be prepared to discuss the limitations of analysis, together with their inherent assumptions and data requirements.

Transport planning works at the interface between technical analysis and policy-making, but very little discussion has developed around the ethical position of planners. It is not sufficient to argue that transport planning is value free and neutral in its approach to analysis. This attitude, deeply rooted in the traditions of positivism, overlooks the value laden assumptions, the quality of the data, the simplification of the decision process, and the inadequacies of both analyst and decision-maker.

The purpose of planning tools is to provide systematic and neutral information to support decision-making, while the ethical content of planning is assumed to be in the definition of the problem and the weighing of information by decision makers. Public hearings and citizen comments on proposals supplement 'objective' analysis by providing personal and subjective views. (Wachs, 1985b, p. xv)

There are no methods which can truly be claimed to be value free and neutral in their approach. Presentation of analysis should acknowledge this situation, with the expert being honest about the limitations of the analysis. Transport planners are both technicians and politicians. On the one hand, transport planning is seen as apolitical with analysis and rationality dominant, whilst on the other hand transport planners see themselves as having a role in bringing about social change. Here, priority must be given to the principles and legitimation of change rather than confidence in the objectivity of the analysis. The code of standards in ethics means adherence to basic values and norms, with a greater openness at all levels of decision-making concerning the advantages and limitations of analysis. This process may weaken the position of the professionals, but by open debate they will receive the respect of the public. This is the choice to be made.

In all situations transport planners should be much more active in presenting and defending their proposals. Increasingly, there is a requirement for partnerships to develop between all the agencies involved with a particular problem (for example, a proposed section of new road). This partnership needs to discuss the technical and planning issues, and the role that a recommended solution might have in addressing those issues. The thinking should not be merely concerned with providing more capacity, but how that investment can achieve multiple objectives, such as also giving priority to public transport, improving the environmental quality, and in giving better access to all. It should also consider the implications for new development and the impact on the existing town centre. This process is a collaborative one that involves all the major actors in a high level debate over the alternatives, so that all perspectives are included and there is a convergence of thinking. It is a consensus seeking process, but it may also result in disagreement. The intention must be to take the technical analysis forwards to the decision-making stage rather than leaving that stage to others. By definition the role of the transport planner is progressing a scheme from the primary concern over the technical issues to a facilitator for discussion, debate and action. Proactive planning is essential if decision-making is to progress towards a more holistic perspective. Planners have much to offer and they should have the confidence to promote defensible solutions to problems. In this way they become more accountable for their actions, and they also gain the respect of the communities in which they work . Planners must be better communicators and advocates for their actions, and become more prepared to promote and protect their stakeholders. In essence, this is the basis of a new contract for partnership, consultation and participation in transport planning.

Infrastructure and finance

Undoubtedly, the most important issue in the next decade concerns the transport infrastructure in Europe, whether it is new infrastructure, an expansion of existing infrastructure, or the reconstruction of existing infrastructure. The basic political questions are the amount of infrastructure which should be provided – whether it should meet unconstrained demand or whether demand should be constrained – and how it should be financed. As described in Chapters 4, 5, 8 and 9, much debate has revolved around the role which the private sector should play. However, most transport projects are large scale, involve a high risk and have long payback periods. It is only where the private sector has a virtual monopoly position (for example, in road bridges across estuaries) that real interest has been raised. Transport projects are notorious for their cost overruns, for technical deficiencies in construction and consequent high maintenance costs, and for optimism in their estimates of future demand – the Channel Tunnel project illustrates all of these problems (Chapter 9).

In the past, the public sector has been the main contributor, but with the increasing costs of new transport projects, and with the desire of governments to reduce public deficits new sources of capital are required. The most obvious means to raise finance would be to charge tolls on existing motorways. It has been estimated (Keating, 1992) that tolls of 5 pence per mile (similar to current toll levels on French autoroutes) on UK motorways would raise about £2 billion per annum and its extension to other roads and to urban areas would raise £6 billion per annum. The key argument is that the tolls must be seen not as a tax but a charge which can be reinvested in the transport system.

The private sector is unlikely to invest in major projects unless the conditions are favourable and a reasonable rate of return can be obtained to satisfy shareholders. This means that low-risk projects would be most desirable so that the following conditions could be met:

- to maintain a virtual monopoly position;
- to meet high levels of existing demand and anticipated growth in that demand;
- to ensure a payback over the medium term (about 10 years);
- to have control over the levels of pricing for the new or upgraded infrastructure;
- to have some guarantee from government of the period over which revenues from tolls will accrue to the private sector.

In most cases, one or more of these conditions are not met and as a consequence the private sector has been reluctant about making any firm commitment. It is in partnership between the private and public sectors where most potential lies. The public sector can anticipate the growth in demand which results from the growth in the economy, rising income levels, and from new developments. It can also assist in the land assembly and the public inquiry processes so that time from project inception to completion is minimized. In certain situations, it can also help with the costs of construction of the actual infrastructure, but in most cases the private sector will manage and run the facility, including setting the levels of tolls or fares. Such an allocation of responsibilities would provide the basis for investment, with the public sector preparing projects for investment. All the necessary statutory processes will have been completed and the project could then be auctioned to the private sector for construction and operation. The actual financial package would depend upon each scheme, with some being most appropriately funded in the public sector and others jointly. A few may even be initiated and funded entirely by the private sector (see discussion in Chapter 9).

There are many possibilities:

◆   through loans from the European Investment Bank and grants from the European Coal and Steel Community (which provided funds for the Channel Tunnel), and the European Community Regional Development Funds and the Transport Infrastructure Fund;

◆   through transport bonds and other long-term investments (using pension funds) – these are extensively used in Japan and are being discussed with respect to the financing of the reconstruction of the London Underground;

◆   through tax incentives to the private sector by making their capital contributions tax deductible;

◆   through tolls and road pricing, as is in operation in Singapore, in several Norwegian cities, in the operation of the Birmingham Northern Relief Road (from 2004), and being proposed in Central London (from 2003);

◆   through shadow tolls;

◆   through employment taxes (as in Paris and other French cities) or a tax on petrol (as in Germany and the United States).

The previous impasse between the public and private sectors has now been overcome, and there is a strong convergence in both thinking and practice. The distinction between the two sectors is becoming less clear and in 10 years time there may be just an investment sector through which funds are channelled. At present, both sectors still have an important role to play in the construction, renewal and maintenance of the road, rail and air transport infrastructure. Yet it should be noted that the public sector cannot withdraw completely, as much of the funding will still remain the responsibility of government. But the role for transport planning must be to facilitate private sector and joint ventures through advice, predictions, land assembly and accelerated public inquiry procedures. The public sector still has the key role to play in a range of different situations:

♦ in engaging interest groups and the general public in planning and decision-making throughout the process;
♦ where the market fails and intervention takes place for accessibility, distributional and equity reasons;
♦ where there are significant externalities involving the use of non-renewable resources, land acquisition, safety and environmental concerns;
♦ where transport interacts with other sectors, such as the generation effects of new developments and priorities given for regional or local development objectives;
♦ where transport has national and international implications, such as promoting London as a world city or maintaining high-quality international air and rail links.

The lessons from the last few decades are clear, namely that under-investment in the transport infrastructure may be politically expedient, but it is a disaster for travellers in the future. We are now living with the consequences of that under-investment and the situation will get substantially worse before we see an improvement, even if the levels of funding are committed on the scale proposed in the Transport 2010 document (DETR, 2000b). This under-investment cuts across all forms of transport and may have been instrumental in contributing to the deterioration in the quality of life. This in turn has encouraged movement out of cities and ironically greater car dependence and more (and longer distance) travel. There must be a commitment from government to maintaining a desirable level of investment in transport, allocated as a proportion of GDP. At present about 1.1% of the EU's GDP is allocated to investment in the transport infrastructure (divided 65% to road, 25% to rail and 10% to other), and this level has declined from about 1.6% of GDP in 1990 (EU, 2000). The UK level is currently below that average (0.8%) and this in no way makes up for the under-investment of the previous 30 years.

The future leisure society

Throughout this book, links have been drawn between transport and other sectors, in particular the planning and development processes. In addition, the focus has been on placing the UK experience in the wider European and worldwide context of the demographic and technological changes taking place in society. The late twentieth century witnessed the second great economic transformation (Castells, 1990). The first was the switch from agrarian to an industrial mode of production, and this in turn led to the development of transport and the high mobility, car based society. The second has been the transition from an industrial mode of production to an information mode where the fundamental inputs are knowledge based.

Castells argues that the three major changes in the capitalist economy in the 1980s were:

- ◆ an increasing surplus from production with a substantial growth in profits;
- ◆ a change in the role of the state away from political legitimation and social redistribution towards political domination and capital accumulation;
- ◆ accelerated internationalization.

The net result of these trends has been a sharp spatial division of labour, the decentralization of particular production functions, flexibility in location, and the generation of information innovation milieux. Space has become increasingly irrelevant as cities have spread and many routine information functions have centred on peripheral, low-cost and low-density sites. The innovation milieux are elite locations where high technology activities are centred, and these locations form the magnets for economic growth and dynamism (Hall, 1988). It is here that the global shifts in power have taken place, and it is through the control of technology that power will be maintained (Toffler, 1991).

What then are likely to be the key concerns for transport planners in the twenty-first century? Some issues have been highlighted in Chapters 8 and 9. The globalization of the capitalist economies will accelerate, and this will not just involve the countries of the West and Japan, but those of Central and Eastern Europe, together with other emerging economies on the Pacific rim on both the Asian and American sides. There will be a transition from centrally controlled and planned economies to market based economies. It is here that a clear understanding is required of both the links between economic growth and transport demand and between the operation of the monetary economy and transport market.

Technological change will fundamentally influence the location of economic development and the function and form of cities. Convergence of computing and communications will allow a user friendly interface for many transactions, for information, and for business and social activities. As Hall (1991) states, the potential for such a system is vast as

it would allow almost infinite dispersion of informational activities across nations and continents, including the spread of home-work and telecommuting. (p. 19)

But equally, information and technology may cause divergence as knowledge to use the system is not universal, as costs of access to the high-quality broad-band communications systems will be expensive, and as control of these systems may reside in the hands of a few multinational companies. Power may be concentrated in existing centres of information exchange at accessible points on the network or in the few world cities where rapid innovation and high-level service competition can take place. Second-order cities and those in peripheral locations will continue to be centres of low demand and low innovation.

Similar developments can be seen in transport where key centres in Europe are likely to be located on the high-speed rail network, particularly at interchanges between road and rail, and between road, rail and air. The international nodal points will be located where the global airlines have their hub operations, and where good quality transport links are available to the local and national centres of population and activity – technopoles. Increasingly, these nodal points will also be centres of the international information networks – the logistics platforms.

As with the industrial revolution, the technological revolution is likely to promote concentration, but at an international rather than national level. Smaller national centres in the hierarchy will be important, but at the national and local levels. It is difficult to speculate on the exact locations of the new global and national city hierarchy. However, the demand for transport will increase in scale, in range, and in quality as this globalization takes place.

Technology may also be socially divisive. In the past there has been a clear division between rich and poor, both within individual countries and between states. With trade and technological barriers being dismantled, those differences are being emphasized with new migrations to the rich countries, but also with a new entrepreneurship in some rapidly developing economies. However, technology may also act as a barrier with a division between those who are technologically rich and those who are not. It will not just be financial wealth that is important, but also technological knowledge.

The increased polarization of the population between those who are financially secure and technologically rich, and those who are not, may lead to political and social instability, particularly if this trend is reflected in the job market through the need for new skills or through the casualization of work. These mega trends, together with the ageing of the population, will lead to an increase in non-work related activities, particularly leisure with many people having time and resources to spend. Others will have resources but less time, and some will have time but no resources. In the UK, 41% of people took no holidays in 1998 (defined as lasting over 4 days), and this proportion has remained constant for the last 30 years. But those taking two or more holidays has risen from 15% in 1971 to 25% in 1998 (GSS, 2000).

New approaches are needed to investigate the burgeoning leisure market and the growth in air travel. In a real sense, these two sectors (leisure and air) encapsulate some of the key questions still to be resolved:

The limits to travel – Growth rates of 5–6% per annum in air travel, mean a doubling of demand every 12 years. The leisure and air businesses markets can cope with such an increase and they would probably relish the challenge. But what is the impact of such growth in terms of the quality of the environment that people are looking for in the leisure activities, and what are the costs in terms of energy consumption and emissions of CO_2 and NO_x? Can the global environment accommodate such rates of growth in consumption? There are no easy answers to this challenge and it reflects many of the same debates that were apparent in the 1960s and 1970s with respect to road transport. It raises again the arguments about whether planners should meet demand by increasing supply, or whether demand management should be introduced in air transport through limits on capacity and substantially raising prices so that the externalities are paid for by the traveller.

Not all travel is derived demand – The traditional view that travel is only undertaken because of the benefits derived at the destination is no longer generally applicable. Substantial amounts of leisure travel are undertaken for their own sake and the activity of travelling is valued. This conclusion has enormous implications for transport analysis as most conventional analysis is based on the premise that travel distances should be short and that travel time should be minimized. If this no longer holds for the majority of travel (and leisure related travel is increasing, accounting for over 50% of current travel), then much of conventional transport 'wisdom' needs to be reassessed.

Necessity to take an intermodal perspective on transport – There seems to be considerable potential within Europe to reduce short distance (energy and pollution intensive) travel by air through the linking of the air and rail networks, as has been achieved in Charles de Gaulle and Frankfurt airports. Through innovative hubbing solutions, it is possible to optimize the use of air space and rail systems. This applies to passenger travel and high value freight movements. However, if such a move merely increases the capacity of the system overall, then short distance air will be replaced by long distance air. This may help the capacity problem for airlines, but it may increase the environmental costs of transport. The key issue here is that decisions should be made that are holistic and consider the implications across all modes and sectors.

Sustainable development through sustainable tourism – Much analysis describes the growth in tourism, but there is much less appreciation of the negative impacts that overuse has on particular locations. Escape theory (Banister, 2000*b* and Heinze, 2000) argues that leisure mobility is an attempt to compensate for declining quality of life. It makes a powerful case for the growth in leisure activities and tourism, but the increasing pressures also destroy the very quality that people want. The question here is how can appropriate limits be set, and how can policy instruments be used to limit numbers of tourists to particular locations. Time management offers one interesting set of possibilities through peak demand spreading and encouraging flexibility in use. Again there are parallels between the new leisure activities and adjusting the existing patterns of demand to take account of sustainable development priorities.

Sustainable supply chains – There is an increasing awareness worldwide of the transport costs entailed in providing goods to the final user. One of the main growth sectors in air travel has been the freight sector where high value goods are carried long distances to give year round availability (for example, for fruit). Two issues arise here, one is to ensure the full environmental costs are paid and the EU and national governments are beginning to tackle the advantageous position of the airlines (for example, in terms of imposing tax on kerosene). The other is to give the consumer full information on the transport content of goods, but also to describe sourcing conditions so that environmental and social quality conditions can be met. Again, some industries (for example, The Body Shop) are responding positively to this challenge.

Impact of technology and flexibility – The new technology provides tremendous opportunity and choice in leisure activities, whether this means time spent in the home at the terminal or out of the home at local or national facilities. The knowledge base is extended and this may result in more travel, but more important is the transfer of power from the producer to the customer. Increasingly, users will control their leisure activities tailored to their own specific requirements (at a price). Consumers will determine what type of leisure activity they participate in, where and when it takes place, and who actually goes with them – and the range of alternatives will also increase substantially.

There will never be a shortage of challenges for transport planning. It seems at last that the 'straightjacket' of the systems approach has been loosened, and the full range of new methods and approaches are being embraced. This is good news as the new policy agenda requires the utilization of the full range of available methods to unravel the complexity of current travel patterns. Many of the basic underlying premises on which much of transport planning was built have now been questioned, and it is likely that a 'new look' transport planning will at last emerge that balances the political, economic, social and environmental constraints under which public policy decisions are made. We are in interesting times!

Abbati, C. degli (1986) *Transport and European Integration*. Brussels: Commission of the European Communities, European Perspectives.

Adams, J. (1981) *Transport Planning: Vision and Practice*. London: Routledge and Kegan Paul.

Alexander, E.R. (1984) After rationality what? *Journal of the American Planning Association*, **50**(1), pp. 62–69.

Altshuler, A. (1965) *The City Planning Process*. Ithaca, NY: Cornell University Press.

Altshuler, A. (1979) *The Urban Transportation System: Politics and Policy Innovation*, Cambridge Mass: MIT Press.

Ambrose, P. (1986) *Whatever Happened to Planning*. London: Methuen.

Anas, A., Arnott, R. and Small, K.A. (1998) Urban spatial structure. *Journal of Economic Literature*, **36**, pp. 1426–1464.

Andersen Consulting (1990) *The Impact of Environmental Issues on Business – A Guide for Senior Management*. London: Andersen Consulting.

Annema, J.A. and Van Wee, B. (2000) Transport and the greenhouse effect. The role of research in Kyoto-related climate policy in the Netherlands, *European Journal of Transport and Infrastructure Research*, 0(0), pp. 57–70 (trial version, November).

Apel, D. and Pharoah, T. (1995) *Transport Concepts in European Cities*. Aldershot: Avebury.

Appleyard, D. (1981) *Liveable Streets*. Berkeley and Los Angeles: University of California Press.

Association of County Councils (ACC) (1991) Towards a Sustainable Transport Policy. Paper prepared by the Association of County Councils, London.

Association of European Airlines (1987) *Capacity of Aviation Systems in Europe: Scenario on Airport Congestion*. Brussels: AEA.

Atkins, S. (1987) The crisis for transportation planning modelling. *Transport Reviews*, 7(4), pp. 307–325.

Atkins, S. (1990) Personal security as a transport issue: A state of the art review. *Transport Reviews*, **10**(2), pp. 111–124.

Atkins, S. (1992) TDM, AQD's, AVR's, SLO's and TMA's: An Introduction to Transportation Planning in Southern California. Paper presented at the Universities Transport Studies Group Conference, Newcastle upon Tyne.

Axhausen, K. (2002) Definition of movement and activity for transport modelling, in Hensher, D.A. and Button, K. (eds.) *Handbooks in Transport: Transport Modelling*. Oxford: Elsevier, and from www.ivt.baug.ethz.ch/veroeffent-arbeitsbericht.html.

Ayres, R.U. (1998) *Turning Point: The End of the Growth Paradigm*. London: Earthscan.

Bagley, M.N., Mannering, F. and Mokhtarian, P. (1994) Telecommuting Centers and Related Concepts. University of California Institute of Transportation Studies, Berkeley CA, access@uclink.berkeley.edu.

Banister, D. (1977) Car Availability and Usage: A Modal Split Model based on these Concepts. University of Reading, Geographical Paper No 58.

Banister, D. (1978) The influence of habit formation on modal choice – a heuristic model. *Transportation*, **7**(1), pp. 5–18.

Banister, D. (1981) The response of the Shire Counties to the question of transport needs. *Traffic Engineering and Control*, **23**(10), pp. 488–491.

Banister, D. (1983) Transport needs in rural areas – A review and proposal. *Transport Reviews*, **3**(1), pp. 35–49.

Banister, D. (1985) Deregulating the bus industry in Britain – the proposals. *Transport Reviews*, **5**(2), pp. 99–103.

Banister, D. (1989) Congestion: Market pricing for parking, *Built Environment*, **15**(3/4), pp. 251–256.

Banister, D. (1990) Privatisation in transport: From the company state to the contract state, in Simmie, J. and King, R. (eds.) *The State in Action: Public Policy and Politics*. London: Pinter, pp. 95–116.

Banister, D. (1992*a*) Energy use, transport and settlement patterns, in Breheny, M. (ed.) *Sustainable Development and Urban Form*. London: Pion, pp. 160–181.

Banister, D. (1992*b*) The British Experience of Bus Deregulation in Urban Transport: Lessons for Europe. University College London, Planning and Development Research Centre, Working Paper 5.

Banister, D. (1992*c*) Demographic Structure and Social Behaviour. Proceedings of the 12th International Symposium on Theory and Practice in Transport Economics – Transport Growth in Question. Lisbon, pp. 109–150.

Banister, D. (1993) Policy responses in the UK, in Banister, D. and Button, K. (eds.) *Transport, the Environment and Sustainable Development.* London: Spon, pp. 53–78.

Banister, D. (1997*a*) Reducing the need to travel. *Environment and Planning B*, **24**(3), pp. 437–449.

Banister, D. (1997*b*) Transport and air quality in the UK. *Greener Management International – Special Issue on Air Quality*, **20**, Winter, pp. 90–96.

Banister, D. (ed.) (1998) *Transport Policy and the Environment.* London: Spon.

Banister, D. (1999) Planning more to travel less: Land use and transport. *Town Planning Review*, **70**(3), pp. 313–338.

Banister, D. (2000*a*) Sustainable urban development and transport: A Eurovision for 2020. *Transport Reviews*, **20**(1), pp 113–130.

Banister, D. (2000*b*) The tip of the iceberg: Leisure and air travel. *Built Environment*, **26**(3), pp. 226–235.

Banister, D. (ed.) (1989) The Final Gridlock. *Built Environment*, **15**(3/4), pp. 159–256.

Banister, D. and Bayliss, D. (1991) Structural Changes in Population and Impact on Passenger Transport Demand. European Conference of Ministers of Transport, Round Table 88. Paris, pp. 103–142.

Banister, D. and Botham, R. (1985) Joint land use and transport planning: The case of Merseyside, in Harrison, A. and Gretton, J. (eds.) Transport UK1985: *An Economic, Social and Policy Audit.* Newbury: Policy Journals, pp. 95–100.

Banister, D. and Norton, F. (1988) The role of the voluntary sector in the provision of rural services – the case of transport. *Journal of Rural Studies*, **4**(1), pp. 57–71.

Banister, D. and Mackett, R. (1990) The minibus: Theory and experience, and their implications. *Transport Reviews*, **10**(3), pp. 189–214.

Banister, D. and Pickup, L. (1990) Bus transport in the Metropolitan Areas and London, in Bell, P. and Cloke, P. (eds.) *Deregulation and Transport Market Forces in the Modern World.* London: Fulton, pp. 67–83.

Banister, D. and Berechman, J. (2000) *Transport Investment and Economic Development.* London: UCL Press.

Banister, D. and Berechman, J. (2001) Transport investment and the promotion of economic growth. *Journal of Transport Geography*, **9**(3), pp 209–218.

Banister, D. and Marshall S. (2000) *Encouraging Transport Alternatives: Good Practice in Reducing Travel.* London: The Stationery Office.

Banister, D.J., Bould, M. and Warren, G. (1984) Towards needs based transport planning. *Traffic Engineering and Control*, **25**(7/8), pp. 372–375.

Banister, D., Berechman, J. and De Rus, G. (1992) Competitive regimes within the European bus industry: Theory and practice. *Transportation Research*, **26A**(2), pp. 167–178.

Banister, D., Cullen, I. and Mackett, R. (1991) The impacts of land use on travel demand, in Rickard, J. and Larkinson, J. (eds.) *Longer Term Issues in Transport*. Aldershot: Avebury, pp. 81–130.

Banister, D., Andersen, B. and Barrett, S. (1995) Private sector investment in transport infrastructure in Europe, in Banister, D., Capello, R. and Nijkamp, P. (eds.) *European Transport and Communications Networks: Policy Evolution and Change*. London: Belhaven, pp 191–220.

Banister, D., Gérardin, B. and Viegas, J. (1998) Partnerships and responsibilities in transport: European and urban policy priorities, in Button, K., Nijkamp, P. and Priemus, H. (eds.) *Transport Networks in Europe: Concepts, Analysis and Policies*. London: Edward Elgar, pp 202–233.

Banister, D., Stead, D., Steen, P., Dreborg, K., Åkerman, J., Nijkamp, P. and Schleicher-Tappeser, R. (2000) *European Transport Policy and Sustainable Mobility*. London: Spon Press.

Batey, P.W. and Breheny, M.J. (1978) Methods in strategic planning. *Town Planning Review*, **49**, pp. 259–273, and pp. 502–518.

Batty, M. (1976) *Urban Modelling: Algorithms, Calibrations and Predictions*. Cambridge: Cambridge University Press.

Batty, M. (1981) A perspective on urban systems analysis, in Banister, D. and Hall, P. (eds.) *Transport and Public Policy Planning*. London: Mansell, pp. 421–440.

Batty, M. (1989) Urban modelling and planning: Reflections retrodictions and prescriptions, in Macmillan, B. (ed.) *Remodelling Geography*. Oxford: Basil Blackwell, pp. 147–169.

Beauregard, R. (1990) Bringing the city back in. *Journal of the American Planning Association*, **56**(2), pp. 210–215.

Becker, H.A. (1997) *Social Impact Assessment*. London: UCL Press, Social Research Today 10.

Beesley, M. and Kain, J.F. (1964) Urban form, car ownership and public policy: An appraisal of *Traffic in Towns*. *Urban Studies*, **1**(2), pp. 174–203.

Beesley, M. and Kain, J.F. (1965) Forecasting car ownership and use. *Urban Studies*, **2**(2), pp. 163–185.

Bendixson, T. (1989) Transport in the Nineties: The Shaping of Europe. A Study commissioned by the Royal Institution of Chartered Surveyors, London.

Benwell, M. (1980) The Contribution of the Social Sciences to Transport Research in France. Transport and Road Research Laboratory, SR 637.

Berlinski, D. (1976) *On Systems Analysis: An Essay Concerning the Limitations of Mathematical Methods in the Social, Political and Biological Sciences*. Cambridge, Mass: MIT Press.

Bernard, M.J. (1981) Problems in predicting market response to new transportation technology, in Stopher, P.R., Meyburg, A.H. and Brög, W. (eds.) *New Horizons in Travel Behaviour Research*. New York: Lexington Books, pp. 465–87.

Bertolini, L. (1998) Station area redevelopment in five European countries: An international perspective on a complex planning challenge. *International Planning Studies*, 3(2), pp 163–184.

Blowers, A. (1986) Town planning – paradoxes and prospects. *The Planner*, April, pp. 82–96.

Bluestone, B. and Harrison, B. (1982) *The Deindustrialization of America: Plant Closures, Community Abandonment, and the Dismantling of Basic Industry*. New York: Basic Books.

BMA (1997) *Road Transport and Health*, Report by the British Medical Association. London: BMA.

Boddy, M. and Thrift, N. (1990) Socioeconomic Restructuring and changes in Patterns of Long Distance Commuting in the M4 Corridor. Final report to the ESRC.

Bonnafous, A. (1987) The regional impact of the TGV. *Transportation*, 14(2), pp. 127–137.

Bontje, L. and Jolles, A. (2000) Amsterdam – The Major Projects. Amsterdam: Drukkerij Mart. Spruijt BV.

Boyce, D., Day, N. and McDonald, C. (1970) *Metropolitan Plan Making*, Monograph Series 4. Philadelphia: Regional Science Research Institute.

Breheny, M. (1989) The planning paradox: Strategic issues and pragmatic responses in the south of England. *Built Environment*, 14(3/4), pp. 225–239.

Breheny, M. (1991) The renaissance of strategic planning? *Environment and Planning B*, 18(3), pp. 233–249.

Breheny, M. and Roberts, A.J. (1981) Forecasts in structure plans. *The Planner*. July/August, pp. 102–103.

Brewer, G.D. (1973) *Politicians, Bureaucrats and the Consultant: A Critique of Urban Problem Solving*. New York: Basic Books.

Bristow, A. and Nellthorp, J. (2000) Transport project appraisal in the European Union. *Transport Policy*, 7(1), pp 51–60.

British Road Federation (1972) *Basic Road Statistics*. London: British Road Federation.

Brög, W. (1992) Structural Changes in Population and Impact on Passenger Transport – Germany. ECMT Round Table 88. Paris, pp. 5–42.

Brotchie, J.F., Dickey, J.W. and Sharpe, R. (1980) *TOPAZ – General Planning Technique and Its Application at Regional, Urban and Facility Planning Levels*. Berlin: Springer-Verlag.

Bruton, M.J. (1985) *Introduction to Transportation Planning*. London: UCL Press, reprinted 1992.

Burnett, P. and Hanson, S. (1982) The analysis of travel as an example of complex human behaviour in spatially-constrained situations: Definitions and measurement issues. *Transportation Research*, 16A(2), pp. 87–102.

Butler, E.W., Chapin, F.S., Hemmens, G., Kaiser, E.J., Steginan, M.A. and Weiss, S.F. (1969) Moving Behavior and Residential Choice – A national survey. Highway Research Board, Special Report 81.

Button, K. (1991) Transport and communications, in Rickard, J. and Larkinson, J. (eds.) *Longer Terms Issues in Transport*. Aldershot: Avebury, pp. 323–360.

CAG Consultants (2001) State of Sustainable Development in the UK: Central/Local Government Focus, Report to the Sustainable Development Commission, Final Report, February.

Cairns, S., Hass-Klau, C. and Goodwin, P. (1998) *Traffic Impact of Highway Capacity Reductions: Assessment of the Evidence*. London: Landor Publishing.

California Transportation Commission (1987) *California Congestion: Its Effect Now and in the Future*. Sacramento, CA: California Transportation Commission.

Carlstein, T., Parkes, D. and Thrift, N. (eds.) (1978) *Timing Space and Spacing Time*, Volume 2, *Human Activity and Time Geography*. London: Edward Arnold.

Castells, M. (1983) *The City and the Grassroots: A Cross-Cultural Theory of Urban Social Movements*. London: Edward Arnold.

Castells, M. (1990) *The Informational City: Information, Technology, Economic Restructuring and the Urban Regional Process*. Oxford: Blackwell.

CEBR (1994) Roads and Jobs: The Economic Impact of Different Levels of Expenditure on the Roads Programme, Report by the Centre for Economics and Business Research Ltd for the British Road Federation, June.

Central Statistical Office (1997) *Social Trends*, Vol. 27. London: The Stationery Office.

Cervero, R. (1989) *America's Suburban Centers: The Land-Use Transportation Link*. Boston: Unwin Hyman.

Cervero, R. and Hall, P. (1990) Containing traffic congestion in America. *Built Environment*, 15(3/4), pp. 176–184.

Cervero, R. and Landis, J. (1995) Development impacts of urban transport: A US perspective, in Banister, D. (ed) *Transport and Urban Development*. London: Spon, pp 136–156.

Champion, A.G., Green, A.E., Owen, D.W., Ellis, D.J. and Coombes, M.G. (1987) *Changing Places: Britain's Demographic, Economic and Social Complexion*. London: Edward Arnold.

Chen, K. (1990) Driver information systems: A North American perspective. *Transportation*, 17(3), pp. 251–62.

Chicago Area Transportation Study (1960) *Final Report*. Chicago.

Civil Aviation Authority (1989) *Traffic Distribution Policy for Airports Serving the London Area*. CAP 559. London: Civil Aviation Authority.

Coburn, T.M., Beesley, M.E. and Reynolds, D.J. (1960) The London-Birmingham Motorway: Traffic and Economics. Road Research Laboratory, Department of Scientific and Industrial Research, Technical Paper 46. London: HMSO.

Coindet, J.-P. (1988) Financing Urban Public Transport in France. Paper presented at the PTRC Annual Conference, University of Sussex.

Commission for Integrated Transport (1999) National Road Traffic Targets, Report for CfIT, November, www.cfit.gov.uk/reports/nrtt99/index.htm.

Commission of the European Communities (CEC) (1985) *Official Journal of the European Communities*, C 144, 13 June.

Commission of the European Communities (CEC) (1990) *Green Paper on The Urban Environment*. EUR 12902. Brussels: CEC.

Commission of the European Communities (CEC) (1991) *The DRIVE Programme in 1991*. DGXIII DRIVE 1202. Brussels: CEC.

Commission of the European Communities (CEC) (1992*a*) *Green Paper on The Impact of Transport on the Environment A Community Strategy for Sustainable Development*. DGVII. Brussels: CEC.

Commission of the European Communities (CEC) (1992*b*) *The Future*

Development of the Common Transport Policy. Com (92)494. Brussels: CEC.

Commission of the European Communities (CEC) (1993) Towards Sustainability: A European Community Programme of Policy and Action in Relation to the Environment and Sustainable Development (Fifth Environmental Action Plan), CEC, Brussels.

Commission of the European Communities (CEC) (1995) *The Citizens' Network*. European Commission Green Paper (COM (95) 601). Brussels: CEC.

Commission of the European Communities (CEC) (1998*a*) Memorandum on the Policy Guidelines of the White Paper on a Common Transport Policy, 18 July, Brussels.

Commission of the European Communities (CEC) (1998*b*) *The Common Transport Policy – Sustainable Mobility: Perspectives for the Future*. Communication from the Commission to the Council, the European Parliament, the Economic and Social Committee and the Committee of the Regions, Brussels, December, COM (1998) 716 Final. Brussels: CEC.

Commission of the European Communities (CEC) (2001) European Transport Policy for 2010: Time to Decide. The White Paper on Transport Policy, Brussels, 12 September 2001, http://europa.eu.int/comm/energy/ transport/en/lb en.html.

Committee on the Medical Effects of Air Pollutants (1998) *The Quantification of the Effects of Air Pollution on Health in the United Kingdom*. London: The Stationery Office.

Coombe, R.D., Forshaw, I.G. and Bamford, T.J.G. (1990) Assessment in the London Assessment Studies. *Traffic Engineering and Control*, 31(10), pp. 510–518.

Coopers and Lybrand (1990) *Internationale vergelijking infrastructuur*. Rotterdam: Coopers and Lybrand.

Coopers and Lybrand, Deloitte (1990) *The Environmental White Paper: This Common Inheritance, Briefing Digest*. London: Coopers and Lybrand, Deloitte.

Council for the Protection of Rural England (2001) *Running to Stand Still? An Analysis of the 10 Year Plan for Transport*. London: CPRE.

Creighton, R. L. (1970) *Urban Transportation Planning*. Chicago: University of Illinois Press.

Cullinane, S. and Stokes, G. (1998) *Rural Transport Policy*. Oxford: Pergamon.

Cullingworth, B. (1988) *Town and Country Planning in Britain*. London: Unwin Hyman.

Cullingworth, B. (1999) *Planning in the USA: Policies, Issues and Processes.* London: Routledge.

Dahl, R. (1972) *Der Anfang von Ende des Autos* (The Beginning of the End of the Car). Munden: Langeswiesche Brandt.

Daly, A. (1981) Behavioural travel modelling: Some European experience, in Banister, D. and Hall, P. (eds.) *Transport and Public Policy Planning.* London: Mansell, pp. 307–316.

David Simmonds Consultancy (1998) Review of Land Use/Transport Interaction Models, Report to SACTRA.

Davidoff, P. (1965) Advocacy and pluralism in planning. *Journal of the American Planning Association*, **31**(4), pp. 331–338.

Deakin, E. (1990*a*) Land use and transportation planning in response to congestion problems: A review and critique. *Transportation Research Record*, No.1237, pp. 77–86.

Deakin, E. (1990*b*) Toll roads: A new direction for US highways? *Built Environment*, **15**(3/4), pp. 185–194.

Department of Energy (1990) *Energy Use and Energy Efficiency in UK Transport up to the Year 2010.* Energy Efficiency Office, Energy Efficiency Series 10. London: HMSO.

Department of the Environment (1972*a*) *Getting the Best Roads for Our Money: The COBA Method of Appraisal.* London: HMSO.

Department of the Environment (1972*b*) *New Roads in Towns.* London: HMSO.

Department of the Environment (1973) *Greater London Development Plan: Report of the Panel of Inquiry.* London: HMSO.

Department of the Environment (1975) *Transport Supplementary Grant Submission for 1977/78.* Circular 125/75. London: HMSO.

Department of the Environment (1983) *Streamlining the Cities: Government Proposals for Reorganising Local Government in Greater London and the Metropolitan Counties.* Cmnd 9063. London: HMSO.

Department of the Environment (1986) *The Future of Development Plans: A Consultation Paper.* London: HMSO.

Department of the Environment (1989) *The Future of Development Plans.* Cm 569. London: HMSO.

Department of the Environment (1990) *This Common Inheritance: Britain's Environmental Strategy.* Cm 1200. London: HMSO.

Department of the Environment (1992) *Climate Change: Our National Programme for CO_2 Emissions.* A Discussion Document. London: Department of the Environment.

Department of the Environment (1994) *Sustainable Development: The UK Strategy*, London: HMSO, Cm 2426, January.

Department of the Environment (1996) *Code of Practice for Major Planning Inquiries*, DOE Circular 5/96. London: HMSO.

Department of the Environment and Department of Transport (1994) *Planning Policy Guidance: Transport* (PPG 13). London: HMSO.

Department of the Environment, Transport and the Regions (1997*a*) *National Travel Survey*. London: HMSO, Table 2.12.

Department of the Environment, Transport and the Regions (1997*b*) *Building Partnerships for Prosperity*, Cm 3814. London: The Stationery Office.

Department of the Environment, Transport and the Regions (1997*c*) *National Road Traffic Forecasts (Great Britain)*. London : HMSO.

Department of the Environment, Transport and the Regions (1997*d*) *Land Use Change in England*, No. 12. Department of the Environment, Transport and the Regions. London: DETR.

Department of the Environment, Transport and the Regions (1997*e*) *Transport Statistics Great Britain 1997*. London: The Stationery Office.

Department of the Environment, Transport and the Regions (1997*f*) *Digest of Environmental Statistics 1997*. London: The Stationery Office.

Department of the Environment, Transport and the Regions (1997*g*). *Air Quality and Traffic Management* [LAQM. G3(97)]. London: The Stationery Office.

Department of the Environment, Transport and the Regions (1998*a*) *A New Deal for Transport: Better for Everyone*. London: The Stationery Office.

Department of the Environment, Transport and the Regions (1998*b*) A New Deal for Trunk Roads in England: Guidance on the New Approaches to Appraisal, London: The Stationery Office.

Department of the Environment, Transport and the Regions (1998*c*) *Modernising Planning – A Policy Statement by the Minister for the Region, Regeneration and Planning*. London: DETR.

Department of the Environment, Transport and the Regions (1999*a*) *A Better Quality of Life: A Strategy for Sustainable Development for the UK*. London: DETR, www.sustainable-development.gov.uk/.

Department of the Environment, Transport and the Regions (1999*b*) *Supplementary Guidance to the Regional Development Agencies*. London: DETR.

Department of the Environment, Transport and the Regions (1999*c*) *Annual Report 1999–2000*. London: DETR, www.detr.gov.uk/annual1999/19.html.

Department of the Environment, Transport and the Regions (2000*a*) *Transport Statistics Great Britain 2000 Edition*. London: The Stationery Office.

Department of the Environment, Transport and the Regions (2000*b*) *Transport 2010: The Ten Year Plan*. London: DETR.

Department of the Environment, Transport and the Regions (2000*c*) *Our Towns and Cities: The Future – Delivering an Urban Renaissance*. London: DETR.

Department of the Environment, Transport and the Regions (2000*d*) *Our Countryside: The Future. A Fair Deal for Rural England*. London: DETR.

Department of the Environment, Transport and the Regions (2000*e*) *Tracking Congestion and Pollution, the Government's First Report under the Road Traffic Reduction* (National Targets) Act 1998. London: DETR.

Department of the Environment, Transport and the Regions (2001) *Revised PPG13 on Transport*. London: The Stationery Office.

Department of Trade (1974) *Maplin: Review of Airport Project*. London: HMSO.

Department of Trade and Industry (1997) *Digest of UK Energy Statistics*. London: The Stationery Office.

Department of Transport (1976*a*) *Transport Policy: A Consultation Document*, Vols. I and II. London: HMSO.

Department of Transport (1976*b*) *Report on the Location of Major Inter-Urban Road Schemes with Regard to Noise and Other Environmental Issues* (Chairman: J. Jefferson). London: HMSO.

Department of Transport (1977*a*) *Transport Policy:* White Paper, Cmnd 6836. London: HMSO.

Department of Transport (1977*b*) *Report of the Advisory Committee on Trunk Road Assessment* (Chairman: Sir George Leitch). London: HMSO.

Department of Transport (1978*a*) *Policy for Roads: England 1978*. Cmnd 7132. London: HMSO.

Department of Transport (1978*b*) *Report on the Review of Highway Inquiry Procedures*. Cmnd 7133. London: HMSO.

Department of Transport (1979*a*) *Trunk Road Proposals – A Comprehensive Framework for Appraisal* (Chairman: Sir George Leitch). London: HMSO.

Department of Transport (1979*b*) *Road Haulage Operators' Licensing*. Report of the Independent Committee of Inquiry (Chairman: Sir C. Foster). London: HMSO.

Department of Transport (1980) *Report of the Inquiry into Lorries, People and the Environment* (Chairman: Sir A. Armitage). London: HMSO.

Department of Transport (1983) *Manual of Environmental Appraisal*. London: HMSO.

Department of Transport (1984) *Buses*. White Paper, Cmnd 9300. London: HMSO.

Department of Transport (1986*a*) *Urban Road Appraisal Report of the Standing Advisory Committee on Trunk Road Appraisal*. (Chairman: Professor T. E. H. Williams). London: HMSO.

Department of Transport (1986*b*) *The Government Response to the SACTRA Report on Urban Road Appraisal*. London: HMSO.

Department of Transport (1988) *National Travel Survey 1985/86*. London: HMSO.

Department of Transport (1989*a*) *National Road Traffic Forecasts (Great Britain)*. London: HMSO.

Department of Transport (1989*b*) *Roads for Prosperity*. Cm 693. London: HMSO.

Department of Transport (1989*c*) *New Roads by New Means*. Cm 698. London: HMSO.

Department of Transport (1990) *Transport Statistics Great Britain 1979–1989*. London: HMSO.

Department of Transport (1991*a*) *Report 1991: The Government's Expenditure Plans 1991–92 to 1993–94*. Cm 1507. London: HMSO.

Department of Transport (1991*b*) *Bus and Coach Statistics: Great Britain, 1990/91*. London: HMSO.

Department of Transport (1996) Transport – The Way Forward. The Government's Response to the Great Transport Debate, London: HMSO, Cm 3234.

Department of Transport (1992*a*) *Transport Statistics Great Britain 1992*. London: HMSO.

Department of Transport (1992*b*) *Assessing the Environmental Impact of Road Schemes*. Report of the Standing Committee on Trunk Road Appraisal. London: HMSO.

Departments of the Environment and Department of Transport (1996) Local Authority Circular on Air Quality and Traffic Management Consultation Paper, London, December.

Dix, M. (1981) Contributions of psychology to developments in travel demand modelling, in Banister, D. and Hall, P. (eds.) *Transport and Public Policy Planning*. London: Mansell, pp. 369–386.

Dollinger, H. (1972) *Die Totale Autogesellschaft* (The Total Car Society). Passau: Carl Hanser.

Domencich, T. and McFadden, D. (1975) *Urban Travel Demand: A Behavioural Analysis*. Amsterdam: North Holland.

Downs, A. (1992) *Stuck in Traffic: Coping with Peak-Hour Traffic Congestion*. Washington DC : Brookings Institute.

Dreborg, K. (1996) Essence of backcasting. *Futures*, **28**(9), pp. 813–828.

Duncan, S.S. (1990) Do housing prices rise that much? A dissenting view. *Housing Studies*, **5**(3), pp. 195–208.

Dunleavy, P. and Duncan, K. (1989) Understanding the Politics of Transport. Paper presented at an ESRC Seminar on Policy Making Processes in Transport, University College London.

Dunleavy, P. and O'Leary, B. (1987) *Theories of the State: The Politics of Liberal Democracy*. London: Macmillan.

Dunn, J.A. (1998) *Driving Forces: The Automobile, Its Enemies and the Politics of Mobility*. Washington: The Brookings Institute.

Dupuit, J. (1844) On the measurement of the utility of public work. *Annales des Ponts et Chaussees*, 2nd Series, Vol.8. (Translation: Barback, R.H. (1952) *International Economic Papers*, **2**, pp. 83–110.)

Dupuy, G. (1978) *Une technique de planification au service de l'automobile*. Paris: Centre à la Recherche d'Urbanisme.

EC Council Directive (1985) On the assessment of the effects of certain public and private projects on the environment. *Official Journal of the European Communities*, L175, pp. 40–48.

EC DRIVE II BATT Consortium (1992) Initial ideas on defining objectives. Internal Working Note 1. University College London.

Echenique, M. H., Flowerdew, A. D. J., Hunt, J. D., Mayo, T. R., Skidmore, I. J. and Simmonds, D. C. (1990) The MEPLAN models for Bilbao, Leeds and Dortmund. *Transport Reviews*, **10**(4), pp. 309–322.

Economist Intelligence Unit (1964) Urban transport planning in the United Kingdom. *Motor Business*, **40**, pp. 17–32.

Ellickson, R. C. (1977) Suburban growth controls: An economic and legal analysis. *The Yale Law Journal*, **86**(2), pp. 385–511.

Else, P. and Trinder, E. (1991) Transport priorities in local government. *Traffic Engineering and Control*, **32**(5), pp. 261–263.

Engwicht, D. (1992) *Towards an Eco-City: Calming the Traffic*. Sydney: Envirobook.

Environment and Transport Planning (1991) *Traffic Calming Manual*.

Unpublished Consultants Report. Brighton: ETP.

Ettema, D.F. and Timmermans, H. (Eds) (1997) *Activity Based Approaches to Travel Analysis*. Oxford: Pergamon.

Etzioni, A. (1967) Mixed scanning: A third approach to decision making. *Public Administration Review*, Winter, pp. 217–242.

European Conference of Ministers of Transport (1982) *Review of Demand Models*. Report of the ECMT Round Table 58. Economic Research Centre. Paris: OECD.

European Conference of Ministers of Transport (1992) *High Speed Rail*. Report of the ECMT Round Table 87. Economic Research Centre. Paris: OECD.

European Conference of Ministers of Transport (2000) *Strategic Environmental Assessment*. Paris: EMCT/OECD.

European Investment Bank (1991) *Briefing Series: Communications*. Luxembourg: EIB.

European Parliament (1991) Community Policy on Transport Infrastructures. European Parliament, Research and Development Papers Series on Regional Policy and Transport 16. Luxembourg.

European Round Table of Industrialists (1988) *Need for Renewing Transport Infrastructure in Europe. Proposals for Improving the Decision Making Process*. Brussels: ERTI.

European Round Table of Industrialists (1990) *Missing Networks in Europe. Proposals for the Renewal of Europe's Infrastructure*. Brussels: ERTI.

European Union (2000) *EU Transport in Figures*, Statistical Pocketbook, DG Energy and Transport in cooperation with Eurostat. Luxembourg: EU.

Evans, S.H. and Mackinder, I.H. (1980) Predictive Accuracy of British Transport Studies. Paper presented at the PTRC Annual Conference. University of Warwick.

Fainstein, S.S. (1991) Promoting economic development: Urban planning in the United States and Great Britain. *Journal of the American Planning Association*, 57(1), pp. 22–33.

Faludi, A. (1987) *A Decision Centred View of Environmental Planning*. Oxford: Pergamon.

Faludi, A. and Van der Valk, A. (1994) *Rule and Order: Dutch Planning Doctrine in the Twentieth Century*. Amsterdam: Kluwer Academic Publishers.

Federal Highway Administration (1988) America's Challenge for Highway Transportation in the 21st Century. Interim Report on the Future National Highway Program Taskforce.

Federal Highway Authority (FHWA) (1984) *Highway Statistics*. Washington DC: US Department of Transportation.

Ferreira, L. J. (1981) Long Term Urban Transportation Planning Methodology in France: A Review. Institute for Transport Studies, University of Leeds, WP 151.

Fletcher, T. and McMichael, A. (1997) (eds.) *Health at the Crossroads: Transport Policy and Urban Health*. Chichester: John Wiley and Sons.

Flyvbjerg, B. (1998) *Rationality and Power*. Chicago: University of Chicago Press.

Foglesong, R. E. (1986) *Planning the Capitalist City*. Princeton, NJ: Princeton University Press.

Foot, D. (1981) *Operational Urban Models: An Introduction*. London: Methuen.

Forester, J. (1980) Critical theory and planning practice. *Journal of the American Planning Association*, **46**(2), pp. 275–286.

Forrester, J. W. (1969) *Urban Dynamics - City Growth, Stagnation and Decay*. Cambridge, Mass: MIT Press.

Foster, C. D. (1974) *Lessons of Maplin: Is the Machinery of Government Decision- Making at Fault?* London: Institute of Economic Affairs.

✓ Foster, C. D. and Beesley, M. E. (1963) Estimating the social benefit of constructing an underground railway in London. *Journal of the Royal Statistical Society A*, **126**(1), pp. 46–92.

Frank Graham Consulting Engineers (1991) *Royal Borough of Kensington and Chelsea Traffic Calming Study*. Hertford: Frank Graham Consulting Engineers.

Freeman Fox, Wilbur Smith and Associates (1968) *West Midlands Transport Study*. Report for Local Authorities, Ministry of Transport, Ministry of Housing, and Local Transport Organizations.

Friedmann, J. (1973) *Retracking America: A Theory of Transactive Planning*. Garden City: Doubleday.

Gakenheimer, R. (1976) *Transportation Planning as a Response to Controversy: The Boston Case*. Cambridge, Mass: MIT Press.

Gent, H.A. and Nijkamp, P. (1989) Devolution of Transport Policy in Europe. Report produced as part of the ESF-NECTAR activities. Amsterdam.

Gérardin, B. (1989*a*) *Possibilities for, and Costs of; Private and Public Investment in Transport*. European Conference of Ministers of Transport, Round Table 80. Paris: OECD

Gérardin, B. (1989*b*) Financing the European High Speed Rail Network. Paper presented at the PTRC Annual Conference, University of Sussex.

Gérardin, B (1990) *Private and Public Investment in Transport: Possibilities and Costs.* European Conference of Ministers of Transport, Round Table 81, Paris, Economic Research Centre. Paris: OECD, pp. 5–32.

Gérardin, B (1993) Financing infrastructures and transport systems in Central and Eastern Europe, submission by the European Conference of Ministers of Transport to the 2nd Pan-European Conference on Transport, CEMT/CS(93)43, Paris, September.

Gérardin, B (1995) *Financing Infrastructures and Transport Systems in Central and Eastern Europe.* Report to the Council of Ministers of the European Conference of Ministers of Transport. Paris: OECD.

Gerardin, B. and Viegas, J. (1992) European Transport Infrastructure and Networks: Current Policies and Trends. Paper presented at the NECTAR International Symposium, Amsterdam.

Gillespie, A. and Cornford, J. (1995) Network diversity or network fragmentation? The evolution of European telecommunications in competitive environments, in Banister, D. Capello, R. and Nijkamp, P. (eds.) *European Transport and Communications Networks.* London: Wiley, pp. 319–332.

Girnau, G. (1983) Wo Kann Gespart Werden im U- und Stadtbahnbau. *Der Nahverkehr*, 1/83, pp. 8–16.

Glaister, S. (1990) Bus Deregulation in the UK. Draft paper presented to the World Bank, p. 36.

Goddard, J. (1989) *Urban Development in an Information Economy: Policy Issues.* OECD Group on Urban Affairs, Project on Urban Impacts of Technological and Socio-Demographic Change, UP/TSDC (89)1. Paris: OECD.

Goldner, W. (1971) The Lowry model heritage. *Journal of the American Institute of Planners*, 37(1), pp. 100–110.

Golob, T.F. (1999) A simultaneous model of household activity participation and trip chain generation. *Transportation Research*, 34B(5), pp. 355–376.

Golob, T.F. and McNally M.G. (1997) A model of activity participation and travel interactions between household heads. *Transportation Research*, 31B(3), pp. 177–191.

Gomez Ibanez, J. and Meyer, J. (1995) Private toll roads in the United States: Recent experiences and prospects, in Banister, D. (ed.) *Transport and Urban Development.* London: Spon, pp. 248–271.

Goodwin, P. (1990) Demographic impacts, social consequences, and the transport policy debate. *Oxford Review of Economic Policy*, 6(2), pp. 76–90.

Goodwin, P. (1991) The right tools for the job: A research agenda, in Rickard, J. and Larlanson, J. (eds.) *Longer Term Issues in Transport*. Aldershot: Avebury, pp. 1–40.

Goodwin, P. and Layzell, A. (1985) Longitudinal analysis for public transport policy issues, in Jansen, G. R. M., Nijkamp, P. and Ruijgrok, C. J. (eds.) *Transportation and Mobility in an Era of Transition*. Amsterdam: North Holland, Elsevier, pp. 185–200.

Goodwin, P., Hallett, S., Kenny, F. and Stokes, G. (1991)Transport: The New Realism. Report to the Rees Jeffreys Road Fund. Oxford.

Gordon, P. and Richardson, H. W. (1989) Gasoline consumption and cities: A reply. *Journal of the American Planning Association*, 55(3), pp. 342–346.

Government Statistical Services (2000) *Social Trends 2001*, GSS, London: The Stationery Office.

Grandjean, A. and Henry, C. (1984) Economic rationality in the development of a motorway network. *Transport Reviews*, 4(2), pp. 143–158.

Grant-Muller, S., Mackie, P., Nellthorp, J. and Pearman, A. (2001) Economic appraisal of European transport projects: The state of the art revisited. *Transport Reviews*, 21(2), pp. 237–262.

Greater London Authority (2001) *Transport Strategy for London*. London: GLA, June.

Greene D. L. (1987) Long run vehicle travel prediction from demographic trends. *Transportation Research Record*, No. 1135, pp. 1–9.

Grieco, M., Pickup, L. and Whipp, R. (eds.) (1989) *Gender, Transport and Employment The Impact of Travel Constraints*. Aldershot: Avebury.

Group Transport 2000 Plus (1990) Transport in a Fast Changing Europe. Report produced by the Group set up by Karel Van Miert, Transport Commissioner of the European Commission.

Gwilliam, K.M. (1990) A European perspective: Long term research needs in the Netherlands, in Rickard, J. and Larkinson, J. (eds.) *Longer Term Issues in Transport*. Aldershot: Avebury, pp. 245–304.

Hagerstrand, T. (1970) What about people in regional science? *Papers of the Regional Science Association*, 24(1), pp. 7–21.

Hall, P. (1980) *Great Planning Disasters*. London: Weidenfeld and Nicholson.

Hall, P. (1985) Urban transportation: Paradoxes for the 1980s, in Jansen G.R. M., Nijkamp, P. and Ruijgrok, C.J. (eds.) *Transportation and Mobility in an Era of Transition*. Amsterdam: Elsevier, North Holland, pp. 367–375.

Hall, P. (1988) *Cities of Tomorrow: An Intellectual History of Urban Planning and Design in the Twentieth Century*. Oxford: Blackwell.

Hall, P. (1989) The turbulent eighth decade: challenger to American city planning. *Journal of the American Planning Association*, 55(2), pp. 275–282.

Hall, P. (1991) Three systems, three separate paths. *Journal of the American Planning Association*. 57(1), pp. 16–20.

Hall, P. and Hass-Klau, C. (1985) *Can Rail Save The City?* Farnborough: Gower.

Hall, P. and Markusan, A. (1985) *Silicon Landscapes*. London: George Allen and Unwin.

Hall, P. and Pfeiffer, U. (2000) *Urban Future 21 – A Global Agenda for Twenty First Century Cities*, London: Spon Press.

Hamnett, C. (1999) *Winners and Losers: Home Ownership in Modern Britain*, London: UCL Press.

Hamnett, C., Harmer, M. and Williams, P. (1991) *Safe as Houses: Housing Inheritance in Britain*. London: Paul Chapman Publishing.

Handy, C. (1995) *The Age of Unreason*. London : Arrow.

Hanson, M.E. (1992) Automobile subsidies and land use: Estimates and policy responses. *Journal of the American Planning Association*, 58(1), pp. 60–71.

Harvey, D. (1978) On planning the ideology of planning, in Burchall, R. W. and Sternlieb, G. (eds.) *Planning Theory in the 1980s*. New Brunswick, NJ: Centre for Policy Research, Rutgers University.

Hass-Klau, C. (1990) *The Pedestrian and City Traffic*. London: Beihaven.

Hay, A. (1992) Equity in transport planning? *The Planner*, 78(7), pp. 12–13.

Hay, A. and Trinder, E. (1991) Concepts of equity, fairness and justice expressed by local transport policy makers. *Environment and Planning C*, 9(4), pp. 453–465.

Healey, P. (1997) *Collaborative Planning: Shaping Places in Fragmented Societies*. Basingstoke: Macmillan.

Heggie, I. (1978) Putting behaviour into behavioural models of travel choice. *Journal of the Operational Research Society*, 29(6), pp. 541–550.

Heinze, G.W. (2000) *Transport and Leisure*. Report of the ECMT Round Table 111 on Transport and Leisure, European Research Centre. Paris: OECD, pp. 1–51.

Hemmens, G.C. (1968) Survey of planning agency experience with urban development models, data processing and computers. 5R97. Highway Research Board, Washington DC.

Hensher, D.A. (1989) Behavioural and resource values of travel time savings: A bicentennial update. *Australian Road Research*, **19**(3), pp. 223–229.

Hepworth, M. and Ducatel, K. (1992) *Transport in the Information Age: Wheels and Wires*. London: Belhaven.

Hey, C. (1996) The incorporation of the environmental dimension into the Transport Policies of the EU, EURES Discussion Paper No. 50. Institute for Regional Studies in Europe, Freiburg.

Hillman, M., Adams, J. and Whitelegg, J. (1990) *One False Move . . . A Study of Children's Independent Mobility*. London: Policy Studies Institute.

Hills, M. (1973) *Planning for Multiple Objectives*. Monograph No 5. Philadelphia PA: Regional Science Research Institute.

HM Treasury (1989) *Government Plans*. Cm 621. London: HMSO.

Hollatz, J, W. and Tamms, F. (eds.) (1965)*Die Kommunalen Verkehrsprobleme in der Bundesrepublik Deutschland* (The Traffic Problems of Local Authorities in the Federal Republic of Germany). Essen: Vulkan.

Holliday, I., Marcou, G. and Vickerman, R. (1991) *The Channel Tunnel: Public Policy, Regional Development and European Integration*. London: Belhaven.

Hoos, I. R. (1972) *Systems Analysis in Public Policy: A Critique*. Berkeley, CA: University of California Press.

House of Commons (1990) *Roads for the Future*. Report of the Transport Committee. Session 1989–1990, HC 198, Vols. I and II. London: HMSO.

House of Commons (1999) *Transport White Paper, Comments from the Environment, Transport and Regional Affairs Select Committee on the Transport White Paper*, Ninth Report. London: The Stationery Office, April.

House of Commons Expenditure Committee (1973) *Urban Transport Planning*, Vols. I, II, III. London: HMSO.

House of Commons Treasury Committee (2000) *Private Finance Initiative*, 4th Report of the Session 1999-2000, Number 147. London : The Stationery Office.

House of Lords Select Committee on Science and Technology (1987) *Innovation in Surface Transport*. HL 57-I. London: HMSO.

Hutchinson, B. (1974) *Principles of Urban Transport Systems Planning*. New York: McGraw Hill.

Hutchinson, B. (1981) Urban transport policy and policy analysis methods. *Transport Reviews*, **1**(2), pp. 169–188.

Innes, J. (1995) Planning theory's emerging paradigm: Communicative action and interactive practice. *Journal of Planning Education and Research*, **14**(3), pp. 183–189.

Institute of Civil Engineers (1983) *Report of the Civil Engineering Task Force on Public Inquiry Procedures* (Chairman: Lord Vaizey). London: Institute of Civil Engineers.

Institute of Transportation Engineers (1980) Evaluation of the accuracy of past urban transportation forecasts. *ITE Journal*, February, pp. 24–34.

Jones, D., May, A. and Wenban-Smith, A. (1990) Integrated transport studies: Lessons from the Birmingham Study. *Traffic Engineering and Control*, **31**(11), pp. 572–576.

Jones, P. (1990) (ed.) *New Developments in Dynamic and Activity Based Approaches to Travel Analysis*. Aldershot: Gower.

Jones, P., Clarke, M. and Dix, M. (1983) *Understanding Travel Behaviour*. Aldershot: Gower.

Kain, J. F. (1978) The use of computer simulation models for policy analysis. *Urban Analysis*, **5**(2), pp. 175–189.

Kashima, S. (1989) Advanced traffic information systems in Tokyo. *Built Environment*, **15**(3/4), pp. 244–250.

Kawashima, H. (1990) Japanese perspective of driver information systems. *Transportation*, **17**(3), pp. 263–84.

Kay, J.A. and Thompson, D. (1986) *Privatisation and Regulation – The UK Experience*. Oxford: Oxford University Press.

Keating, G. (1992) Toll tales on the highway to prosperity. *Independent*, 16 December.

Khattak, A.J., Schofer, J.L. and Koppelman, F.S. (1991) Commuters' Enroute Diversion and Return Decisions: IVHS Design Implications. Proceedings of the 6th International Conference on Travel Behaviour, Quebec, pp. 362–76.

Kitamura, R. (1990) Panel analysis in transportation planning: an overview. Transportation Research, **24A**(6), pp. 401–415.

Knack, R. (1995) BART'S village vision. *Planning*, **61**(1), pp. 18–21.

Kostyniuk, L.P. and Kitamura, R. (1987) Effects of aging and motorization on travel behavior: *An Exploration. Transportation Research Record*, No. 1135, pp. 31–36.

Kroes, E.P. and Sheldon, R.J. (1988) Stated preference methods. An introduction. *Journal of Transport Economics and Policy*, **22**(1), pp. 11–26.

Kroon, M. (1997) Traffic and environmental policy in the Netherlands, in Tolley, R. (ed.) *The Greening of Urban Transport*. Chichester: Wiley, pp. 161–176.

Künzli, N, (2000) Affect on human health of air pollution and its economic cost, Report prepared for the World Health Organisation by the University of Basel, Switzerland, and reported in *The Lancet*, 1 September 2000.

Kutter, E. (1972) *Demographische Determinanten des Städtlischen Personenverkehrs*. Braunschweig: Instituts für Stadtbauwesen, Technische Universität.

Kuznets, S. (1930) *Secular Movements in Production and Prices – the Nature and Bearing on Cyclical Fluctuations*. Boston: Houghton Mifflin.

Lancaster, K.J. (1966) A new approach to consumer theory. *Journal of Political Economy*, **84**(2), pp. 132–157.

Laslett, P. (1990) *A Fresh Map of Life: The Emergence of the Third Age*. London: Weidenfeld and Nicholson.

Lassave, P. and Offner, J.M. (1989) Urban transport: changes in expertise in France in the 1970s and 1980s. *Transport Reviews*, **9**(2), pp. 119–135.

Lave, C. (1990) Things won't get a lot worse: The future of US traffic congestion. University of California, Irvine, Economics Department, Mimeo.

Lawless, P. (1986) *The Evolution of Spatial Policy*. London: Pion.

Lee, D. (1973) Requiem for large scale models. *American Institute of Planners*, **39**(2), pp. 163–178.

Lee, D. (1994) Retrospective on large scale urban models. *Journal of the American Planning Association*, **60**(1), pp. 35–40.

Leibbrand, K. (1964) *Verkehr und Städtebau*. Basel: Birkhauser Verlag.

Lenntorp, B. (1981) A time-geographic approach to transport and public policy planning, in Banister, D. and Hall, P. (eds.) *Transport and Public Policy Planning*. London: Mansell, pp. 387–396.

Levin, P. H. (1979) Highway inquiries: A study in government responsiveness. *Public Administration*, **57**(1), pp. 21–49.

Lex Motoring (1992) Lex Report on Motoring 1992. Report produced by MORI for Lex Motoring, London.

Lichfield, N. (1970) Evaluation methodology of urban and regional plans: A review. *Regional Studies*, **4**(2), pp. 151–165.

Lichfield N. (1992) Planning and markets in transport, in Glaister, S., Lichfield N., Bayliss, D., Travers, T. and Ridley, T.(eds.) *Transport Options for London*. London: London School of Economics, Greater London Group, pp. 142–160.

Lichfield, N. (1996) *Community Impact Evaluation*. London: UCL Press.

Lichfield, N., Kettle, P. and Whitbread, M. (1976) *Evaluation in the Planning Process*. Oxford: Pergamon Press.

Lindblom, C. E. (1959) The science of muddling through. *Public Administration Review*, Spring, pp. 151–169.

Lojkine, J. (1974) *La politique urbaine dans la Region Lyonnaise 1945–72*. Paris: Mouton.

London County Council (1964) *London Traffic Survey*. London: LCC.

London Strategic Policy Unit (1986) Transport Policies for London 1987–88. Transport Policy Group.

Louvière, J. (1988) Conjoint analysis modelling of stated preferences. A review of theory, methods, recent developments and external validity. *Journal of Transport Economics and Policy*, 22(1), pp. 93–120.

Lowry, I. (1964) *A Model of Metropolis*, RM-4035-RC. Santa Monica, California: Rand Corporation.

Lundqvist, L. (1984) Analysing the impacts of energy factors on urban form, in Brotchie, J. F., Newton, P. W., Hall, P. and Nijkamp, P. (eds.) *The Future of Urban Form*. London: Croom Helm.

Mäcke, P. (1964) *Das Prognoseverfahren in der Strassenverkehrsplanung*. Berlin: Bau Verlag.

Mackett, R.L. (1985) Integrated land-use transport models. *Transport Reviews*, 5(4), pp. 325–343.

Mackett, R. L. (1990) Comparative analysis of modelling land-use transport interaction at the micro and macro levels. *Environment and Planning A*, 22, pp. 457–475.

Maddison, D., Pearce, D., Johansson, O., Calthorp, E., Litman, T. and Verhoef, E. (1996) *The True Costs of Road Transport*. London: Earthscan, Blueprint 5.

Maggi, R., Masser, I. and Nijkamp, P. (1992) Missing networks in European transport and communications. *Transport Reviews*, 12(4), pp. 311–321.

Markusan, A. (1985) *Profits, Cycles, Oligopoly and Regional Development*. Cambridge, Mass: MIT Press.

Masser, I., Sviden, O. and Wegener, M. (1992) *The Geography of Europe's Futures*. London: Belhaven.

May, A. (1991) Integrated transport studies: A new approach to urban transport policy formulation in the UK. *Transport Reviews*, 11(3), pp. 223–248.

McCann, J. (1995) Rethinking the economics of location and agglomeration, *Urban Studies*, 32(3), pp. 563–579.

McLoughlin, J.B. (1969) *Urban and Regional Planning: A Systems Approach*. London: Faber and Faber.

Meyer, J.R., Kain, J.F. and Wohl, M. (1965) *The Urban Transportation Problem*. Cambridge, MA: Harvard University Press.

Michelman, F. I. (1967) Property, utility and fairness: Comments on the ethical foundation of 'Just Compensation' law. *Harvard Law Review*, 80(6), pp. 1165–1258.

Miles, I. (1989) The electronic cottage: Myth or reality? *Futures*, 21(1), pp. 47–59.

Ministry of Transport (1963) *Traffic in Towns: A Study of the Long Term Problems of Traffic in Urban Areas* (Chairman: Colin Buchanan). London: HMSO.

Ministry of Transport (1964) *Road Pricing: The Economic and Technical Possibilities* (Chairman: R. Smeed). London: HMSO.

Ministry of Transport (1967) *Better Use of Town Roads*. London: HMSO.

Ministry of Transport (1968) *Traffic and Transport Plans*. Roads Circular 1/68. London: HMSO.

Ministry of Transport (1969) *Roads for the Future: A New Inter-Urban Plan*. London: HMSO.

Ministry of Transport (1970) *Roads for the Future. The New Inter-Urban Plan for England*. Cmnd 4369. London: HMSO.

Mitchell, R. and Rapkin, C. (1954) *Urban Traffic – A Function of Land Use*. New York: Columbia University Press.

Monheim, R. (1980) *Fussgangerbereiche und Fussgangerverkehr in Stadtzentren*. Bonn: Bonner Geographische Abhandlungen.

Monheim, R. (1997a) The evolution from pedestrian areas to 'car-free' city centres in Germany, in Tolley, R. (ed.) *The Greening of Urban Transport*. Chichester: Wiley, pp. 253–266.

Monheim, R. (1997b) Policy issues in promoting the green modes in Germany, in Tolley, R. (ed.) *The Greening of Urban Transport*. Chichester: Wiley, pp. 191–206.

Moss, M.L. (1987) Telecommunications, world cities and urban policy. *Urban Studies*, 24(4), pp. 534–541.

Motte, A. (1997) The institutional relations of plan making, in Healey, P., Khakee, A., Motte, A. and Needham, B. (eds.) *Making Strategic Spatial Plans: Innovation in Europe*. London: UCL Press, pp. 231–254.

MVA (with ITS Leeds and TSU Oxford) (1987) *The Value of Time Savings*. Newbury: Policy Journals.

National Audit Office (1988) *Road Planning*. House of Commons Paper 688. London: HMSO.

National Audit Office (1998) *The Flotation of Rail Track*. Report by the Comptroller and Auditor General HC25 1998/1999. London: NAO.

National Audit Office (2001) Report on the Department of the Environment, Transport and the Regions: The Channel Tunnel Rail Link, House of Commons HC302, London, March.

National Committee on Urban Transportation (1958) *Better Transportation for Your City: A Guide to the Factual Development of Urban Transportation Plans*. Chicago: Public Administration Service.

National Economic Development Council (NEDC) (1991) Amber Alert: Relieving Urban Traffic Congestion, Report of the Traffic Management Systems Working Party, Chaired by John Ashworth.

National Economic Development Office (NEDO) (1984) *A Fairer and Faster Route to Major Road Construction*. London: NEDO.

National Economic Research Associates (1997) Motors or Modems? Virtual Travel becomes a Reality, Report Commissioned by the RAC, London.

Newman, P.W.G. and Kenworthy, J. R. (1989) *Cities and Automobile Dependence*. Aldershot: Gower.

Newman, P.W.G. and Kenworthy, J.R. (1999) *Sustainability and Cities: Overcoming Automobile Dependence*, Washington DC: Island Press.

Nijkamp, P., Rienstra, S. and Vleugel, J. (1998) *Transportation Planning and the Future*. Chichester: Wiley.

Nilles, J. M. (1988) Traffic reduction by telecommunications: A status review and selected bibliography. *Transportation Research*, **22A**(4), pp. 301–317.

Noortman, H. (1988) The changing context of transport and infrastructure policy. *Environment and Planning C*, 6(2), pp. 131–144.

OECD (Organisation for Economic Cooperation and Development) (1988) *Transport and the Environment*. Paris: OECD.

OECD (Organisation for Economic Cooperation and Development) (1989) *Comparative Socio Demographic Trends: Initial Data on Ageing Populations*. OECD Group on Urban Affairs, UP/TSDA(89)1. Paris: OECD.

OECD (Organisation for Economic Cooperation and Development) (1991) *Environmental Indicators*. Paris: OECD.

OECD (Organisation for Economic Cooperation and Development) (1993) *Indicators for the Integration of Environmental Concerns into Transport Policies*, Environmental Monograph No. 80. Paris: OECD.

Orfeuil, J.-P. (1991) Structural Changes in Population and Impact on Passenger Transport Demand. Paper presented at the European Conference of Ministers of Transport, Round Table 88. Paris.

Owen, S. (2001) A Strategic Analysis of the State of Sustainable Development in the UK: The Contribution made by Regional Government, the Private Sector and NGOs. Environmental Resources Management, Oxford, January, http://www.ermuk.com.

Paaswell, R. (1995) ISTEA: Infrastructure investment and land use, in Banister, D. (ed.) *Transport and Urban Development*. London: Spon, pp. 36–58.

Paaswell, R. (2000) Regionalism, planning and strategic investment in the transportation sector, in Beuthe, M. and Nijkamp, P. (eds.) *New Contributions to Transportation Analysis in Europe*. Aldershot: Ashgate, pp. 251–274.

Pack, J.R. (1978) *Urban Models: Diffusion and Policy Applications*, Monograph Series 7. Philadelphia: Regional Science Research Institute.

Pangalos, S. (1989) Prominent Applications and Software. Paper presented at the DRIVE-EUROFRET Workshop on Freight and Fleet Management. Brussels.

Paradeise, C. (1980) *The Contribution of Social Sciences to Transport Research in Germany*. Paris: Institute de Recherche des Transports.

Pas, E.I. and Harvey, A.S. (1997) Time use research and travel demand analysis modelling, in Stopher, P.R. and Lee-Gosselin, M.E. (eds.) *Understanding Travel Behaviour in an Era of Change*. Oxford: Pergamon, pp. 315–338.

Pendyala, R.M., Goulias, K.G. and Kitamura, R. (1991) Impact of Telecommunications on Spatial and Temporal Patterns of Household Travel: An Assessment for the State of California Pilot Projects Participants. Prepared for the California State Department of Transportation, UCD-ITS-RR-91-07.

Plowden, S. (1972) *Towns Against Traffic*. London: Andre Deutsch.

Plowden, W. (1973) *The Motor Car and Politics in Britain*. Harmondsworth: Penguin.

Polak, J. (1987) A comment on Supernak's critique of transport modelling. *Transportation*, **14**(1), pp. 63–72.

✓ Prest, A.R. and Turvey, R. (1965) Cost benefit analysis: A survey. *Economic Journal*, 75, pp. 685–705.

Prevedouros, P. D. and Schofer, J. L. (1989) Suburban transport behaviour as a factor in congestion. *Transportation Research Record*, No. 1237, pp. 47–58.

Price, A. (1999) The new approach to the appraisal of road projects in England. *Journal of Transport Economics and Policy*, 33(2), pp. 221–226.

Priemus, H. and Nijkamp, P. (1992) Randstad policy on infrastructure and transportation: High ambitions, poor results, in Dieleman, E.M. and Musterd, S. (eds.) *The Randstad. A Research and Policy Laboratory.* Amsterdam: Kluwer Academic, pp. 165–191.

Public Accounts Committee (1989) *Road Planning.* 15th Report of the PAC, HC 101. London: HMSO.

Putman, S.H. (1983) *Integrated Urban Models.* London: Pion.

Putman, S.H. (1986) Integrated Policy Analysis of Metropolitan Transportation and Location. Report DOT-P-30-80-32. US Department of Transportation, Washington DC.

Quinet, E. (2000) Evaluation methodologies of transportation projects in France. *Transport Policy,* **7**(1), pp. 27–34.

Ramsey, A. (1997) A systematic approach to planning of urban networks for walking, in Tolley, R. (ed.) *The Greening of Urban Transport.* Chichester: Wiley, pp. 213–228.

RCEP (1994) *Transport and the Environment.* Eighteenth Report of the Royal Commission on Environmental Pollution, Cm 2674. London: HMSO.

RCEP (1997) *Transport and the Environment – Developments since 1994.* Twentieth Report of the Royal Commission on Environmental Pollution, Cm 3752. London: The Stationery Office.

Reade, E. (1987) *British Town and Country Planning.* Milton Keynes: Open University Press.

Rees, R. (1986) Is there an economic case for privatisation? *Public Money,* **5**(4), pp. 19–26.

Reward Group (1990) Annual Survey of Employee Benefits. Mimeo. London.

Richmond, J.E.D. (1990) Introducing philosophical theories to urban transport planning. *Systems Research,* **7**(1), pp. 47–56.

Rickaby, P. (1987) Six settlement patterns compared. *Environment and Planning B,* **14**(3), pp. 193–223.

Rienstra, S (1998) *Options and Barriers for Sustainable Transport Policies: A Scenario Approach.* Rotterdam: Netherlands Economic Institute.

Rietveld, P. (1993) Policy responses in the Netherlands, in Banister, D. and Button, K. (eds.) *Transport, the Environment and Sustainable Development.* London: Spon, pp. 102–113.

Roos, D. and Altshuler, A. (1984) *The Future of the Automobile.* London: George Allen and Unwin.

Rosenbloom, S. and Altshuler, A. (1977) Equity issues in urban transportation. *Policy Studies Journal,* **6**(1), pp. 29–39.

Roskill, Lord Justice (1971) *Report of the Commission on the Third London Airport*. London: HMSO.

Ross, J.E.L. (1998) *Linking Europe: Transport Policies and Politics in the European Union*. London: Praeger.

Roth, G. (1996) *Roads in a Market Economy*. Aldershot: Avebury.

Rothengatter, W. (2000) Evaluation of infrastructure investments in Germany. *Transport Policy*, 7(1), pp. 17–25.

SACTRA (1994) *Trunk Roads and the Generation of Traffic*. Report of the Standing Advisory Committee on Trunk Road Assessment. London: HMSO.

SACTRA (1999) *Transport and the Economy*. Report of the Standing Advisory Committee on Trunk Road Assessment. London: The Stationery Office.

Savitch, H. (1988) *Post Industrial Cities*. Princeton, NJ: Princeton University Press.

Sayer, R.A. (1976) A critique of urban modelling. *Progress in Planning*, 6(3), pp. 187–254.

Schon, D.A. (1983) *The Reflective Practitioner*. New York: Basic Books.

Schumpeter, J.A. (1934) *A Theory of Economic Development*. Cambridge, MA: Harvard University Press.

Scott, M. (1969) *American City Planning Since 1890: A History Commemorating the Fiftieth Anniversary of the American Institute of Planners*. Berkeley: University of California Press.

Self, P. (1975) *Econocrats and the Policy Process: The Politics and Philosophy of Cost-Benefit Analysis*. London: Macmillan.

Self, P. (1980) *Planning the Urban Region*. London: George Allen and Unwin.

Simmonds, D. (2001) A New Look at Multi Modal Modelling: Conclusions and Recommendations. Final report prepared by DSC, ITS Leeds, MVA and John Bates Services for the Department of the Environment, Transport and the Regions, p. 50.

Simpson, B. (1987) *Planning and Public Transport in Great Britain, France and West Germany*. London: Longman.

Skamris, M., and Flyvbjerg, B., (1997) Inaccuracy of traffic forecasts and cost estimates on large transport projects, *Transportation Policy*, 4(3), pp. 141–146.

Smith, P. (1988) *City, State and Market*. Oxford: Blackwell.

Starkie, D.N.M. (1973) Transportation planning and public policy. *Progress in Planning*, 1(4), pp. 313–389.

Starkie, D.N.M. (1982) *The Motorway Age: Road and Traffic Policies in Post-War Britain*. Oxford: Pergamon.

Stigler, G.J. (1975) *The Citizen and the State: Essays on Regulation*. Chicago: University of Chicago Press.

Stopher, P.R. and Lee-Gosselin, M.E. (eds.) (1997) *Understanding Travel Behaviour in an Era of Change*. Oxford: Pergamon.

Supernak, J. (1983) Transportation modelling: lessons from the past and tasks for the future. *Transportation*, **12**(1), pp. 79–90.

Supernak, J. and Stevens, W.R. (1987) Urban transportation modelling: the discussion continues. *Transportation*, **14**(1), pp. 73–82.

Tanaka, M. (1991) Dealing with unanticipated events. *The Wheel Extended*, **76**, p. 30.

Taylor, M.D.P., Young, W., Wigan, M.R. and Ogden, K.W. (1992) Designing a large scale travel demand survey: New challenges and new opportunities. *Transportation Research*, **26A**(3) pp. 247–261.

Tennyson, R. (1991) Interactions between Changing Urban Patterns and Health. Paper presented at the Solar Energy Conference on Architecture in Climate Change, RIBA.

Thomas, K. and Roberts, P. (2000) Metropolitan strategic planning in England: strategies in transition. *Town Planning Review*, **71**(1), pp. 25–49.

Thomson, J.M. (1969) *Motorways in London*. London: Duckworths.

Thomson, J.M. (1974) *Modern Transport Economics*. Harmondsworth: Penguin.

Titheridge, H., Hall, S. and Banister, D. (1999) Sustainable Settlements – A Model for the Estimation of Transport, Energy and Emissions, A Final Report, ESTEEM Working Paper 4, January, p 22, available from h.titheridge@ucl.ac.uk .

Toffler, A. (1991) *Power Shift: Knowledge, Wealth and Violence at the Edge of the 21st Century*. London: Bantam.

Tolley, R. (1990) *Calming Traffic in Residential Areas*. Tregaron, Wales: Brefi Press.

Tolley, R. (ed.) (1997) *The Greening of Urban Transport*. Chichester: Wiley.

Traffic Research Corporation (1969) *Merseyside Area Land-use Transportation Study*. Final Report Vol A.

Transnet (1990) *Energy, Transport and the Environment*. London: Transnet.

Transportation Research Board (TRB) (2001) *Making Transit Work: Insight from Western Europe, Canada and the United States*. Washington DC: TRB, Special Report 257.

Trinder, E., Hay, A., Dignan, J., Else, P. and Skorupski, J. (1991) Concepts of equity, fairness and justice in British Transport legislation 1960–1988. *Environment and Planning C*, **9**(1), pp. 31–50.

Troin, J.F. (1995) *Rail et aménagement du territoire. Des héritages aux nouveaux défis.* Aix-en-Provence: Edisud.

Tyson, W.J. (1988) A Review of the First Year of Bus Deregulation. Report to the Association of Metropolitan Authorities and Passenger Transport Executive Group. Manchester.

Tyson, W.J. (1989) A Review of the Second Year of Bus Deregulation. Report to the Association of Metropolitan Authorities and Passenger Transport Executive Group. Manchester.

Tyson, W.J. (1990) Effects of deregulation on service coordination in the Metropolitan Areas. *Journal of Transport Economics and Policy*, **24**(3), pp. 283–294.

Tyson, W. J. (1991) Urban public transport fare structures, in Rickard, J. and Larkinson, J. (eds.) *Longer Term Issues in Transport.* Aldershot: Avebury, pp. 305–322.

UK Treasury (1984) *Investment Appraisal in the Public Sector: A Technical Guide for Government Departments.* London: HM Treasury.

Urban Task Force (1999) *Towards an Urban Renaissance.* London: The Stationery Office.

US Department of Transportation (1986) Personal travel in the US, in *The Nationwide Personal Transportation Study*, Vol 2. Washington DC: US Department of Transportation.

US Department of Transportation Federal Highways Administration (1968) Instructional Memorandum 50-4-68: Operations Plans for 'Continuing' Urban Transportation Planning. Washington DC.

Van Geenhuizen, M., Van Zuylen, H. and Nijkamp, P. (1998) Limits to predictability, Paper presented at PTRC, September London, and published in *Transportation Planning Methods*, No. 1307, pp. 291–302.

Van Rest, D. (1987) Policies for major roads in urban areas in the UK. *Cities*, **4**(3), pp. 236–252.

Vibe, N. (1991) Are There Limits to Growth in Car Ownership? Some Recent Figures from the Oslo Region. Preliminary paper presented at the ECMT Round Table 88. TP/0371/1991.

Vickerman, R.W. (1991) Transport infrastructure in the European Community: New developments, regional implications and evaluation, in Vickerman, R.W. (ed.) *Infrastructure and Regional Development.* London: Pion, pp. 36–50.

Vigar, G. (2001) *The Politics of Mobility.* London: Spon Press.

Vogel, S. and Rowlands, I. (1990) The challenges and opportunities facing the

European electronics information industry, in Locksley G. (ed.) *The Single European Market and the Information and Communication Technologies*. London: Belhaven.

Voogd, H. (2001) Social dilemmas and the communicative planning paradox. *Town Planning Review*, **72**(1), pp. 77–95.

VROM (1999) Uitvoeringsnota Klimaatbeleid, Deel 1: Binnenlandse maatregelen (Policy document on climate policy. Part 1: domestic measures) Ministry of Housing, Spatial Planning and the Environment (VROM), The Hague (in Dutch).

Vuchic, V.R. (1999) *Transportation for Livable Cities*. Rutgers University, Brunswick NJ: Center for Urban Policy Research Press.

Wachs, M. (1985a) The politicization of transit subsidy policy in America, in Jansen, G. R. M., Nijkamp, P. and Ruijgrok, C J. (eds.) *Transportation and Mobility in an Era of Transition*. Amsterdam: Elsevier, North Holland, pp. 353–366.

Wachs, M. (ed.) (1985b) *Ethics in Planning*. New Brunswick, NJ: Centre for Urban Policy Research, Rutgers University.

Wachs, M. (1990) Regulating traffic by controlling land uses: The Southern California experience. *Transportation*, **16**(3), pp. 241–256.

Wagner, D. and Kleinschmidt, V. (1992) Feasibility of a strategic environmental assessment for the German Federal Traffic infrastructure plan, in Lee, N. and Hughes, P. (eds.) *Strategic Environmental Assessment Legislation and Procedures in the Community*. 2 vols. Brussels: European Commission, pp. 231–250.

Watson, R.T., Ravindranath, N.H., Noble, I.R. and Bolin, B. (eds.) (2001) *Land Use, Land Use Change and Forestry*. Special Report of the Intergovernmental Panel on Climate Change. Cambridge: Cambridge University Press.

Webber, M.M. (1976) *The BART experience – What have we learned?* Monograph 26. Berkeley: Institute of Urban and Regional Development and Institute of Transportation Studies, University of California.

Webster, F.V. and Paulley, N.J. (1990) An international study on land use and transport interaction. *Transport Reviews*, **10**(4), pp. 287–308 and *Transport Reviews*, **11**(3), pp. 197–222.

Webster, M. (1984) *Explanation, Prediction and Planning: The Lowry Model*. Pion: London.

Weiner, E. (1985) Urban transportation planning in the US: An historical overview. *Transport Reviews*, **4**(4), pp. 331–358 and 5(1), pp. 19–48.

Wermuth, M., Kutter, E., Zuinkeller, D., Heidemann, C. and Brog, W. (1990)

Federal Republic of Germany, in Nijkamp, P., Reichman, S. and Wegener, M. (eds.) *Euromobile: Transport, Communications and Mobility in Europe*. Farnborough: Avebury, pp. 185–196.

Whitelegg, J. (1993) *Transport for a Sustainable Future: The Case of Europe*. London : Belhaven.

Whitt, J.A. (1982) *Urban Elites and Mass Transportation: The Dialectics of Power*. Princeton, NJ: Princeton University Press.

Williams, K., Burton, E. and Jenks, M. (eds.) (2000) *Achieving Sustainable Urban Form*. London: Spon Press.

Willson, R. (2001) Assessing communicative rationality as a transportation planning paradigm. *Transportation*, 28(1), pp. 1–31.

Wood, C. and Jones, C. (1991) *Monitoring Environmental Assessment and Planning*. Report by the EIA Centre, Department of Planning and Landscape, University of Manchester, for the Department of the Environment. London: HMSO.

Wood, D. (1992) Environmental quality and value for money, in Banister, D. and Button, K. (eds.) *Transport, the Environment and Sustainable Development*. London: Spon, pp. 212–216.

Wrigley, N. (1986) Quantitative methods: the era of longitudinal data analysis. *Advances in Human Geography*, 10, pp. 84–102.

Yiftachel, O. (1989) Towards a new typology of urban planning theories. *Environment and Planning C*, 16(1), pp. 23–39.

Zopel, C. (1991) A New Transport Policy. Friedrich Ebert Foundation, Economic and Social Policy in Europe, Background Paper No 1. Berlin.